COUNTER EXAMPLES IN DIFFERENTIAL EQUATIONS AND RELATED TOPICS

COUNTER EXAMPLES IN DIFFERENTIAL EQUATIONS AND RELATED TOPICS

John M Rassias

Published by

World Scientific Publishing Co. Pte. Ltd.
P O Box 128, Farrer Road, Singapore 9128
USA office: 687 Hartwell Street, Teaneck, NJ 07666
UK office: 73 Lynton Mead, Totteridge, London N20 8DH

Library of Congress Cataloging-in-Publication data is available.

COUNTER-EXAMPLES IN DIFFERENTIAL EQUATIONS AND RELATED TOPICS

ISBN 981-02-0460-4

Printed in Singapore by JBW Printers & Binders Pte. Ltd.

PREFACE

This book includes various parts of the theory of *Differential Equations and Related Topics* such as: Continuity and Linearity, Differentiability and Analyticity, Extrema, Existence, Uniqueness, Stability, Singularities, Dynamical Systems (e.g.: Regularity, Periodicity, Uniformity, Orientation), Integral Equations (e.g.: Fredholm, Volterra) , Other Topics, and Open Problems.

The present work is a revised and augmented version of a Semester lecture course delivered by me at the School of Topography, G.S.A. , Greece, and the American College of Greece for many academic years.

Differential Equations and Related Topics play an enormously important role in Science, Engineering, and Mathematics. In fact, any phenomenon that changes in a continuous·or quasi continuous manner can be modelled employing *Differential Equations and Related Topics.*

This Collection of Counter-examples is drawn from at least seventy eight related sources (that is, books and journals) for over a period of fifteen years of research work in this field of studies and is designed especially for the students and researchers in Science (e.g.: Physics, Chemistry, Biology, e.t.c.), Engineering, and Applied and Pure Mathematics.

Deep gratitude is due to all those who have generously helped me to carry out this project. My very special thanks to my wife Vassiliki, and daughters Katia and Matina for their understanding. Last, and certainly not least I am grateful to the Consultant Editor Professor J.G. Xu and Publishers of World Scientific for the great cooperation.

John M. Rassias, Ph.D.

CONTENTS

Preface		v
Chapter 1.	**Continuity and Linearity**	1
Chapter 2.	**Differentiability and Analyticity**	4
Chapter 3.	**Extrema**	6
Chapter 4.	**Existence**	10
Chapter 5.	**Uniqueness**	46
Chapter 6.	**Stability**	94
Chapter 7.	**Singularities**	132
Chapter 8.	**Dynamical Systems**	138
Chapter 9.	**Integral Equations**	150
Chapter 10.	**Other Topics**	158
Chapter 11.	**Open Problems**	165
	References	166
	Subject Index	175
	Name Index	181

1. CONTINUITY AND LINEARITY

1.1 *A non-linear function $u = u(x)$, continuous only in an open interval, which at every point x of the interval satisfies the one-dimensional mean value equation*

$$u(x) = \frac{u(x+h) + u(x-h)}{2h} \qquad (*)$$

for corresponding $h = h(x) > 0$.

The following example is due to Max Shiffman (**14**, p. 281–282).

This function is *continuous and piecewise linear in the interval* $0 < x < 1$, and zigzags back and forth between the lines $y = 0$ and $y = 1$. It is clear that it is *not* continuous at points $x = 0$ and $x = 1$.

Assume that the peaks, which lie on the line y=1, have abscissas a_ν. Similarly assume that the peaks, which lie on the line $y = 0$, have abscissas b_ν. Observe that the neighborhood of every point of the interval (0,1), which does *not* coincide with one of the points a_ν or b_ν, u is linear and therefore satisfies equation $(*)$ for some h (in fact, for infinitely many h).

Choose a_ν so that to every a_ν there exist two points $a_\alpha < a_\nu$ and $a_\beta < a_\nu$ for which $a_\nu = \dfrac{a_\alpha + a_\beta}{2}$. Then u satisfies $(*)$ also at peaks on $y = 1$. In fact,

$$h(a_\nu) = a_\beta - a_\nu = a_\nu - a_\alpha (> 0).$$

Similarly choose b_ν so that b_ν does not coincide with a_μ for any ν and μ, and so that in every interval between two adjacent a_ν exactly one b_μ occurs. Then u is non-linear continuous (for $0 < x < 1$) satisfying equation $(*)$.

One of the many ways to construct two sequences a_ν and b_ν

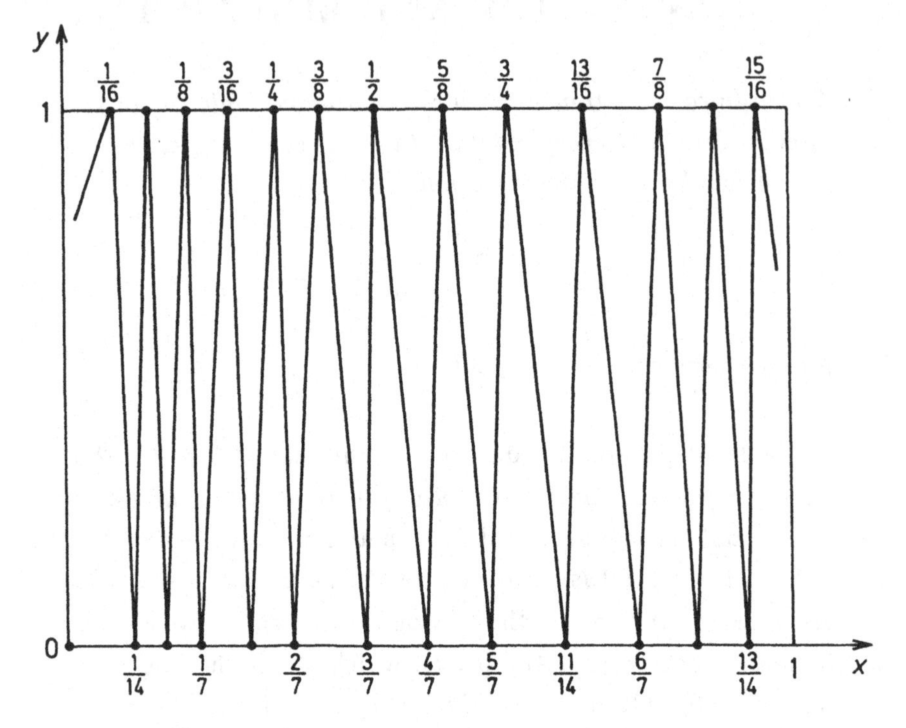

Fig. 1.1 *A non-linear continuous function u satisfying the one-dimensional equation* $u(x) = \frac{u(x+h)+u(x-h)}{2h}$

with the properties described is to *choose* points with abscissas

$$a = \begin{cases} \dfrac{1}{2^k} \quad , \quad \dfrac{3}{2^{k+2}} , \\ 1 - \dfrac{1}{2^k} \quad , \quad 1 - \dfrac{3}{2^{k+2}} \end{cases} \qquad (k = 1, 2, \dots)$$

on $y = 1$, and

$$b = \begin{cases} \dfrac{1}{7}\dfrac{1}{2^{k+1}} \quad , \quad \dfrac{1}{7}\dfrac{3}{2^k} , \\ 1 - \dfrac{1}{7}\dfrac{1}{2^{k-1}} \quad , \quad 1 - \dfrac{1}{7}\dfrac{3}{2^k} \end{cases} \qquad (k = 0, 1, 2, \dots)$$

on $y = 0$, which are symmetric with respect to $x = \dfrac{1}{2}$.

1.2 *A non-linear discontinuous function* $u = u(x)$ *satisfying the one-dimensional mean value equation* $(*)$ *for arbitrary* x *and* $h = h(x) > 0$.

This example is due to G. Hamel (**28**; and **14**, p. 280) and justifies the reason why the continuity of u does not follow directly from the mean value property $(*)$, in general.

1.3 *A discontinuous linear function*

Note that such a function must be very discontinuous. See: B. R. Gelbaum, and J. M. H. Olmsted ("Counterexamples in Analysis", Holden-Day, Inc., San Francisco, p. 33, 1964).

This function is constructed by employing a Hamel basis for the linear space of the reals over the rationals. In fact, this process yields a set $A = \{r_\alpha\}$ of reals r_α so that every real x is expressed uniquely via a linear combination of a finite number of elements of A with rational coefficients q_α ; that is,

$$x = \sum_{i=1}^{k} q_{\alpha_i} r_{\alpha_i} .$$

Therefore the function f is defined by

$$f(x) = \sum_{i=1}^{k} p_{\alpha_i}$$

because of the uniqueness of the representation of x as a linear combination.

It is clear that f is discontinuous because its values are all rational but not all equal. From the definition of f implies its linearity.

2. DIFFERENTIABILITY AND ANALYTICITY

2.1 *A Differential Equation with no power series solution.*

Consider $y' = g(x)$
where

$$g(x) = \begin{cases} e^{-\frac{1}{x^2}} & , \quad x \neq 0 \\ 0 & , \quad x = 0 . \end{cases}$$

See : **22** , p. 724; and B. R. Gelbaum, and J. M. H. Olmsted ("Counterexamples in Analysis" , Holden-Day, Inc., San Francisco, p. 40, 1964).

The function g is infinitely differentiable and so that all of its derivatives at $x = 0$ are equal to 0 ; that is,

$$g(0) = g'(0) = g''(0) = \ldots = g^{(n)}(0) = \ldots = 0 .$$

In fact,

$$\begin{aligned} g'(0) &= \lim_{x \to 0} \frac{g(x) - g(0)}{x - 0} \\ &= \lim_{x \to 0} \frac{\frac{1}{x}}{e^{\frac{1}{x^2}}} = \frac{\infty}{\infty} \\ &= 0 \end{aligned}$$

by De L'Hospital's Rule. Similarly, for the rest of the derivaties at $x = 0$. If g were analytic at $x = 0$, then it would be (at $x = 0$) equal to the power series

$$\begin{aligned} (g(x) =) g(0) + \frac{x}{1!}g'(0) + \frac{x^2}{2!}g''(0) + \ldots + \frac{x^n}{n!}g^{(n)}(0) + \ldots \\ = 0 + \frac{x}{1!}0 + \frac{x^2}{2!}0 + \ldots + \frac{x^n}{n!}0 + \ldots = 0 . \end{aligned}$$

Thus

$$g(x) = 0$$

throughout the interval containing 0, which is a contradiction because only $g(0) = 0$.Therefore the function on g is *not* analytic at 0. Hence the general solution (of the above-mentioned equation)

$$y = \int_0^x g(x)dx + C ,$$

which is valid for all x, is not analytic at 0 either; that is, there is no power series solution.

3. EXTREMA

One-dimensional weak maximum principle:

Let a twice continuously differentiable function $u = u(x)$ at the single variable $x \in I = (a, b)$ be a nonconstant fuction satisfying

$$Lu = u'' + b(x)u' \geq 0 \tag{1}$$

in I, with $b = b(x)$ bounded in I.

Then a nonconstant u can assume its maximum only on ∂I (i.e. $x = a$ or $x = b$).

One dimensional strong maximum principle:

Let $u = u(x)$ be a nonconstant function satisfying (1) in $I = (a, b)$, with $b = b(x)$ bounded in I, and u twice continuously differentiable function at the single variable $x \in I$.

Then u takes its maximum on ∂I and $\frac{du}{dn} > 0$ at such a point, where $\frac{d}{dn}$ denotes the outward derivative on ∂I so that

$$\frac{du}{dn}|_{x=a} = -u'(a), \text{ and } \frac{du}{dn}|_{x=b} = u'(b) . \tag{2}$$

3.1 *An unbounded function $b = b(x)$ for which both one-dimensional weak and strong maximum principles fail to hold in general.*

See : **71**, p. 14-15 ; and M. H. Protter, and H. F. Weinberger ("Maximum Principles in Differential Equations", Prentice - Hall, Englewood Cliffs, N. J., 1967) where there is the *counter-example:*

$$u'' + b(x)u' = 0 \tag{3}$$

with

$$b(x) = \begin{cases} -\dfrac{3}{x} & , \quad x \neq 0 \\ 0 & , \quad x = 0 . \end{cases}$$

It is clear that

$$u = 1 - x^4$$

is a solution of (3) and that the one-dimensiomal weak maximum principle fails to hold on any interval containing $x = 0$ in its interior.

Similarly, if $x = 0$ is an end point, then the one-dimensional strong maximum principle fails to hold, since

$$u'(0) = 0 .$$

Theorem

Consider the linear elliptic equation

$$\sum_{i,j=1}^{n} a_{ij}(x) u_{x_i x_j} + \sum_{i=1}^{n} b_i(x) u_{x_i} + c(x) u = f(x) , \qquad (4)$$

where $x = (x_1, x_2, \ldots, x_n)$, $a_{ij}(x) = a_{ji}(x)$, and the coefficients $a_{ij}(x)$, $b_i(x)$, $c(x)$, and function $f = f(x)$ are defined in an open domain D. If

$$c = c(x) < 0 , \quad \text{and} \quad f = f(x) \geq 0 \text{ (or } \leq 0)$$

hold in D, then the solution $u = u(x)$ of (4) regular in D *cannot* attain a positive maximum (or negative minimum) in D. If

$$c \leq 0 , \quad \text{and} \quad f > 0 \ (or < 0)$$

hold in D, then u *cannot* attain a non-negative maximum (or non-positive minimum) in D.

3.2 *A positive function $c = c(x)$ for which both extremum principles (Theorem above) fail to hold in general.*

Consider equation (**42**, p. 76) :

$$u_{xx} + u_{yy} + 2u = 0 \qquad (4)'$$

where

$$c = 2 \ (> 0) \ .$$

Then the function

$$u = u(x, y) = \sin x \sin y$$

in the square

$$S = \{(x, y) \in \mathbb{R}^2 : 0 \leq x \leq \pi \ , \quad 0 \leq y \leq \pi\}$$

is a solution of $(4)'$ and vanishes on the perimeter of S without vanishing inside S.

3.3 *An unbounded domain D for which both extremum principles (Theorem above) fail to hold in general.*

Consider equation (**42**, p. 184):

$$u_{xx} + u_{yy} - 4u = 0 \tag{4''}$$

in the half plane

$$D \ : \ y \geq 0 \ , \quad \text{and} \quad x \in \mathbb{R} = (-\infty, \infty).$$

This equation $(4)''$ has a regular solution in D which vanishes on $\partial D \ : \ y = 0$, and $x \in \mathbb{R} = (-\infty, \infty)$, without identically vanishing in D.

In fact, such a solution is of the form

$$u = u(x, y) = \operatorname{sh}\left(x\sqrt{2}\right) \cdot \operatorname{sh}\left(y\sqrt{2}\right) \tag{5}$$

Second Counter-Example (**16**, p. 20) :

Consider equation

$$\Delta u \equiv u_{xx} + u_{yy} = 0 \tag{4'''}$$

in the unbounded

$$D = \{(x, y) \in \mathbb{R}^2 \; : \; -\infty < x < \infty \, , \quad 0 < y < \pi\}.$$

This Laplace equation (4)''' has regular solution

$$u = u(x, y) = e^x \; \sin y \tag{6}$$

in D, which vanishes on $\partial D \; : \; y = 0$, and $y = \pi, -\infty < x < \infty$, without identically vanishing in D.

4. EXISTENCE

4.1 *An Initial Value Problem with no solution.*

Consider the differential equation (**75**, p. 144):

$$y'(x) = \begin{cases} \dfrac{1}{x} & , \quad x \neq 0 \\ \\ 0 & , \quad x = 0 \end{cases} \tag{1}$$

with initial condition

$$y(0) = 0 \ . \tag{2}$$

It is clear that Problem (1)-(2) has no solution.

Picard's Theorem

Assume $f = f(x, y)$ a real valued function of the real variables x, y, defined on an open region R of the (x, y)-plane. Let (**a**) f be a continuous function of x, y on R. (**b**) f satisfy *a Lipschitz Condition* with respect to y in R; that is, there is a constant $L \geq 0$ such that for any two points (x, y_1),(x, y_2) of R

$$|f(x, y_1) - f(x, y_2)| \leq L|y_1 - y_2| \tag{3}$$

holds.

Then, for any fixed point (x_0, y_0) of R there is a number $b > 0$ and a function $y = y(x)$ such that

(**i**) $y'(x)$ exists and is continuous in $|x - x_0| \leq b$.

(**ii**) $y'(x) = f(x, y(x))$ for $|x - x_0| \leq b$.

(**iii**) $y(x) = y_0$.

(**iv**) $y(x)$ is the *only* function satisfying conditions (**i**)–(**iii**).

Remark

The character of Picard's Theorem above is *local*; that is, it states only that $b > 0$ exists so that $y = y(x)$ is the solution in $x_0 \leq x \leq x_0 + b$.

4.2 ***An Initial Value problem with local and no global solution***

Consider the differential equation (**75**, p. 150):

$$y'(x) = y^2(x) \tag{4}$$

with initial condition

$$y(0) = 1\,. \tag{5}$$

Then the solution of the initial value Problem (4)-(5) exists in $0 \le x \le 1 - \delta(\delta > 0)$ but not in any interval containing $x = 1$.

In fact,the solution of Problem (4)-(5) is of the form

$$y(x) = \frac{1}{1-x} \tag{6}$$

which is a solution in $0 \le x \le 1 - \delta(\delta > 0)$ but not in any interval containing $x = 1$.

Picard's Theorem provides an *estimate* of $b = \min(a, \frac{c}{M})$, where $M = \max\{|f(x,y)| : (x,y) \in S\}$ and $S = \{(x,y) : |x - x_0| \le a, |y - y_0| \le c\}$.

In particular, here

$$\begin{aligned} M &= \max\{|y^2| : (x,y) \in S\} \\ &= (1+c)^2\,, \end{aligned}$$

and

$$b = \max \min_{a,c>0}\left(a,\ \frac{c}{(1+c)^2}\right).$$

The maximum occurs when

$$a = \frac{c}{(1+c)^2}\,,\quad c \ge 0\,.$$

Hence

$$b = \frac{1}{4}$$

is the best estimate obtained from Picard's Theorem.

Second Counter-Example

Consider equation (4) with initial condition

$$y(1) = -1 \ . \tag{7}$$

The solution of Problem (4)–(7) is

$$y(x) = -\frac{1}{x} \ . \tag{8}$$

However, this solution does *not* at $x = 0$, although $f(x, y) = y^2$ is continuous there (**12**, p. 3).

Remarks

(i) The above-mentioned two counter-examples show that any general existence theorem is necessarily of *local* nature, and existence *in the large* can only be achieved under new conditions on f.

(ii) The solution (8) exists e.g. on (-1,0) but cannot be continued to (-1,0].

Theorem 4.1

Consider the differential equation

$$\frac{dy}{dx} = f(x, y) \tag{9}$$

with initial condition

$$y(x_0) = y_0 \ . \tag{10}$$

Assume that $f = f(x, y)$ and its partial derivative with respect to y are continuous throughout some rectangular region of the (x, y)-plane containing point (x_0, y_0).

Then there is one and only one solution of Problem (9)–(10) on some interval $x_0 - h < x < x_0 + h$.

4.3 *A Line on which Initial Value Problem fails to have a solution.*

Consider (**64**, p. 49–50):

$$f(x,y) = 3xy^{\frac{1}{3}}\ , \tag{11}$$

which is continuous in the entire-plane. However its partial derivative with respect to y,

$$\frac{\partial f}{\partial y} = x \cdot y^{-\frac{2}{3}}\ , \tag{12}$$

fails to exit along the line $y = 0$.

Therefore the function $f = f(x,y)$ above satisfies hypotheses of Theorem 1 in any rectangle that does not contain any part of the x-axis.

Remark:

Function $f = f(x,y)$ fails to be continuous at any point of the xy-plane at which f becomes infinite.

It may happen, however, that the related equation

$$\frac{dx}{dy} = \frac{1}{f(x,y)} \tag{13}$$

does have solution there.

4.4 *A differential equation with no real valued solutions.*

Consider equation (**15**, p. 11):

$$\left(\frac{dy}{dx}\right)^2 + 3 = 0\ . \tag{14}$$

This equation has no real valued solutions because $\frac{dy}{dx}$ is imaginary.

4.5 *A differential system with no solution.*

Consider system (**76**, p. 255,and p. 367):

$$\left.\begin{aligned} \frac{dx}{dt} &= x^2 + y \\ \frac{dy}{dt} &= x + y^2 \end{aligned}\right\} . \tag{15}$$

Each solution

$$x = x(t) \ , \ y = y(t) \tag{16}$$

of (15) satisfying condition

$$x(0) > 0 \ , \ y(0) > 0 \tag{17}$$

cannot exist on an interval of the form $0 \le t < \infty$.

In fact, the trajectory of any solution initiating in the first quadrant is contained in the first quadrant.

Note that

$$\frac{dx(t)}{dt} \ge x^2(t) \ , \ \frac{dy(t)}{dt} \ge y^2(t) \ .$$

4.6 *An Initial Value Problem with no Lipschitz Condition.*

Consider differential equation (**5**, p. 157):

$$\frac{dy}{dx} = e^y \quad (= f(x,y)) \tag{18}$$

with initial condition

$$y(0) = c(\text{: constant}) \tag{19}$$

The ratio

$$\frac{|f(x,y) - f(x,0)|}{|y - 0|} = \frac{e^y - 1}{y} \tag{20}$$

is unbounded if the domain of $e^y = f(x, y)$ is unrestricted.

Note:

The solution of Problem (18)-(19) is

$$y = y(x) = -\ln(e^{-c} - t) \tag{21}$$

and is defined only in the interval

$$-\infty < t < e^{-c}.$$

Remark

There is *no* $\epsilon > 0$ such that equation (18) has a solution defined on all of $t : |t| < \epsilon$ for every initial value; the interval of definition of a solution changes with the initial value.

4.7 *A general differential system with no closed path.*

Consider the autonomous system

$$\left.\begin{aligned} \frac{dx}{dt} &= P(x, y) \\ \frac{dy}{dt} &= Q(x, y) \end{aligned}\right\} \tag{22}$$

in the domain D of the (x, y)-plane, where $P = P(x, y)$ and $Q = Q(x, y)$ have continuous first partial derivatives in D.

Assume that the divergence of $\vec{F} = (P, Q)$:

$$\frac{\partial P}{\partial x} + \frac{\partial Q}{\partial y} (= \operatorname{div} \vec{F} = \nabla \cdot \vec{F}), \tag{23}$$

has the same sign throughout D.

Then system (22) has no closed path in D (**70**, p. 496-497).

In fact, apply *Green's theorem:*

$$\iint_R \left(\frac{\partial P}{\partial x} + \frac{\partial Q}{\partial y} \right) ds = \oint_{\partial R} P dy - Q dx \tag{24}$$

where R is region bounded by closed path ∂R in D. Similarly, if $C = \partial R$ is a closed path of system (22), then

$$\oint_C P dy - Q dx = \int_0^T \left(P\frac{dy}{dt} - Q\frac{dx}{dt} \right) dt$$
$$= \int_0^T (PQ - QP) dt = 0 \ ,$$

where T denotes the period of the solution defining C. Thus

$$\iint_R \nabla \cdot \vec{F} ds = 0 \ , \tag{25}$$

which is a contradiction, because (25) holds only if $\nabla \cdot \vec{F}$ changes sign.

Therefore C is *not* a path of (22) and hence (22) possesses *no* closed path in D.

Note:

The above counter-example is the famous *Bendixson's Nonexistence Criterion.*

4.8 *An Initial Value Problem with no solution on any interval* $[0, \epsilon), \epsilon > 0$.

Consider differential equation

$$\frac{dy}{dx} = \begin{cases} -1 & , \quad y \geq 0 \\ 1 & , \quad y < 0 \end{cases} \tag{26}$$

with initial condition

$$y(0) = 0 \tag{27}$$

It is clear that Problem (26)–(27) has *no* solution on any interval $[0, \epsilon), \epsilon > 0$; a kind of jamming occurs at the endpoint $y = 0$ (**25**, p. 273).

4.9 *A general Cauchy Problem for hyperbolic equations with no solution.*

Consider (**42**, p. 61–62) the partial differential equation

$$u_{xy} = 0 \tag{28}$$

with initial (Cauchy) conditions prescribed on one of the characteristics of (28).

The above-mentioned Cauchy Problem has *no* solution for equation (28), and initial conditions prescribed on one of the characteristics, even when the functions appearing in these conditions are analytic but otherwise arbitrary.

In fact, equation (28) has characteristics the two families of straight lines

$$x = \text{const.}, \qquad \text{and} \qquad y = \text{const.} \tag{29}$$

satisfying the characteristic equation

$$dxdy = 0 \, . \tag{30}$$

Integrating equation (28) with respect to y we obtain

$$u_x(x, y) = \theta(x) \, , \tag{31}$$

where $\theta = \theta(x)$ is an arbitrary function of x possessing the primitive.

Now, integrating relation (31) with respect to x we get

$$u = u(x, y) = \varphi(x) + \psi(y) \, , \tag{32}$$

where $\varphi = \varphi(x)$ is the primitive function of the function $\theta = \theta(x)$ and $\psi = \psi(y)$ is an arbitrary differentiable function of y.

Therefore the general solution of (28) has the form (32).

Assume that the required solution of (28) is an analytic function in the neighborhood of the point $(x_0, 0)$, satisfying initial conditiions

$$\left.\begin{aligned} u(x,0) &= \tau(x) \\ u_y(x,0) &= \nu(x) \end{aligned}\right\} \tag{33}$$

where $\tau = \tau(x)$ and $\nu = \nu(x)$ are analytic in the neighborhood of x_0.

Then since $u = u(x,y)$ has the form (32) there must exist functions $\varphi = \varphi(x)$ and $\psi = \psi(y)$, such that

$$\left.\begin{aligned} &\varphi(x) + \psi(0) = \tau(x) \\ &\psi'(0) = \nu(x) \end{aligned}\right\} \tag{34}$$

This however is possible only when

$$\nu(x) = c(:\ constant)\,. \tag{35}$$

If this condition is satisfied, then function $\psi = \psi(y)$ may be arbitrary and analytic in y so that

$$\psi'(0) = c\ . \tag{36}$$

Thus we choose

$$\varphi(x) = \tau(x) - \psi(0)\ . \tag{37}$$

4.10 *A general Cauchy Problem for elliptic equations with no solution.*

Consider a domain D consisting of two adjacent domains D_1 and D_2 separated by a set S, and of the points of this set, so that

$$D = D_1 \cup D_2 \cup S.$$

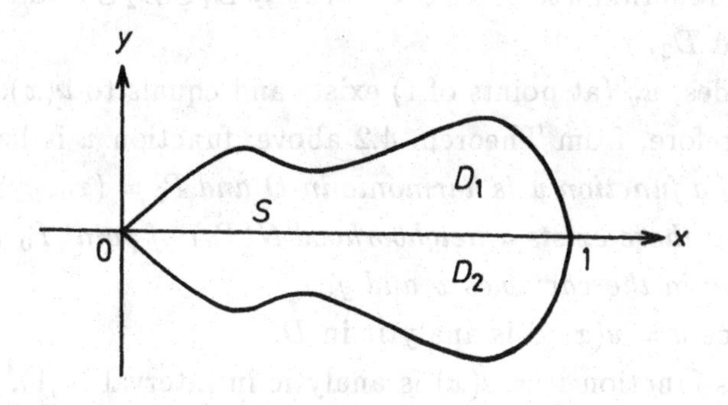

Fig. 4.1

Assume that the closure of S is a surface of class $\mathbf{C}^1$ (**42** p. 215, 235–237).

Theorem 4.2

A function $u = u(x, y)$ of class $\mathbf{C}^1$ in D, harmonic in domains D_1 and D_2 is harmonic in D

Now we'll give the following *counter-example*:

Consider Laplace equation

$$\Delta u \equiv u_{xx} + u_{yy} = 0 \tag{38}$$

with initial (Cauchy) conditions

$$\left.\begin{aligned} u(x,0) &= 0 \\ u_y(x,0) &= \nu(x) \end{aligned}\right\} \tag{39}$$

where $\nu = \nu(x)$ is a function of class $\mathbf{C}^1$ not analytic in the interval (0,1), and $u = u(x, y)$ is harmonic in D_1:$y > 0$, the boundary ∂D_1 of which contains the segment

$$I:\ 0 < x < 1\ , \quad y = 0\ .$$

The function $u = (x, y)$ can be continued into domain D_1 (by Schwarz Reflection Principle), symmetric to D_1 with respect to the x-axis, assuming

$$u(x, y) = -u(x, -y)\ , \quad y < 0\ . \tag{40}$$

This function u is of class $\mathbf{C}^1$ in $D = D_1 \cup D_2 \cup I$, and harmonic in D_1 and D_2.

Besides, u_y (at points of I) exists and equals to $\nu(x)$.

Therefore, from Theorem 4.2 above, function u is harmonic in D. But *if a function u is harmonic in D and $P_0 = (x_0, y_0)$ is a point of D, then there exists a neighborhood $N(P_0)$ of point P_0 in which u is analytic in the variables x and y.*

Hence $u = u(x, y)$ is analytic in D.

Thus function $\nu = \nu(x)$ is analytic in interval (0,1), contradiction, since ν is of class $\mathbf{C}^1$.

4.11 *A nonlinear Cauchy Problem with no solution in the large.*

Consider the nonlinear differential equation

$$u_{xx} = u^2 \tag{41}$$

with initial (Cauchy) conditions

$$\left.\begin{aligned} u(x,-x) &= 6 \\ u_x(x,-x) &= u_y(x,-x) = 12 \end{aligned}\right\} \tag{42}$$

The solution of Problem (41)–(42) becomes infinite on the line

$$x + y = 1\,.$$

Thus this solution does not *not* exist in the large (**23**, p. 120).

4.12 *A general Dirichlet Problem for hyperbolic equations with no solution.*

Consider the wave equation (28) in normal form with boundary values (**23** p. 236):

$$\left.\begin{aligned} u(x,0) &= f(x)\ ,\ 0 \le x \le 1 \\ u(0,y) &= g(y)\ ,\ 0 \le y \le 1 \end{aligned}\right\} \tag{43}$$

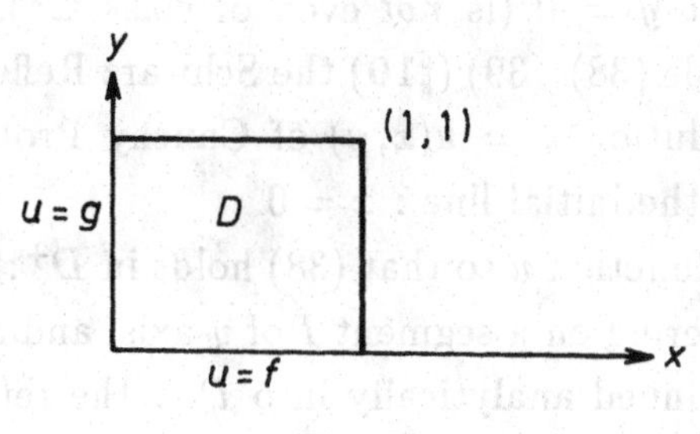

Fig. 4.2

in the unit square $D : 0 < x < 1\ , 0 < y < 1$.

From (43) and the hypothesis that both functions $f = f(x)$, and $g = g(y)$ are smooth we can interpret these values of $u = u(x, y)$ as data for a *characteristic initial value problem* which fixes the solution of equation (28) uniquely throughout the square.

In fact,

$$f(x, y) = f(x) + g(y) \tag{44}$$

when $f(0) = g(0) = 0$.

Therefore the boundary values of u cannot be prescribed arbitrarily on the remaining two sides of D.

Hence a solution u does *not* exist in general when data are imposed on the complete perimeter of the characteristic square, which means that the Dirichlet Problem for equation (28) is *overdetermined* in D.

4.13 *A Cauchy Problem for Laplace equation :* $\Delta u \equiv u_{xx} + u_{yy} = 0$ *and data :* $u(0, y) = 0\ ,\ u_x(0, y) = |y|\ , x > 0$ *with no solution.*

This *counter-example* is similar to (38)–(39). Consider eqation (38) with data:

$$\left.\begin{aligned} u(0, y) &= 0 \\ u_x(0, y) &= |y| \end{aligned}\right\} \tag{39$'$}$$

and $x > 0, -\epsilon < y < \epsilon$. It is clear that

$$\nu(y) = |y|$$

is *not* analytic at $y = 0$ (is *not* even of class $\mathbf{C}^1$). We apply as in counter-example (38)–(39) (♯**10**) the Schwarz Reflection Principle to extend the solution $u = u(x, y)$ of Cauchy Problem (38)–(39)′ backward across the initial line : $x = 0$.

In fact, any function u so that (38) holds in D^+: section of half-plane $x > 0$ bordering on a segment I of y-axis, and such that $u = 0$ on I can be continued analytically into D^-: the reflected image of D^+ with respect to y-axis by the rule (**23**, p. 450–451;**42**, p. 215, 235–237):

$$u(-x, y) = -u(x, y) \ . \tag{40$'$}$$

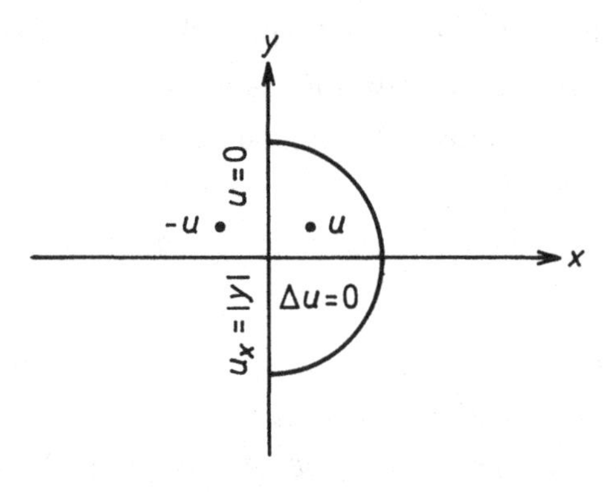

Fig. 4.3

In particular, any solution u of Cauchy Problem (38)–(39)′ has to be regular in a complete neighborhood of each point which lies on I. Thus u_x is regular in I. Hence $\nu = |y|$ has to be analytic in at least some finite interval about origin, contradiction.

4.14 *A Smooth Linear Partial Differential Equation in three independent variables without solution.*

See:**23**, p. 452–455; **47**, p. 155–158 and **19**, p. 138–139.

Consider the first order equation

$$\left[\left(\frac{\partial}{\partial x} + i\frac{\partial}{\partial y}\right) - 2i(x+iy)\frac{\partial}{\partial t}\right]u = g'(t) \tag{45}$$

where $g' = g'(t)$ is a given real function of t and *fails* to be analytic at $t = 0$.

Then equation (45) has *no* continuously differentiable solution in any neighborhood of the origin.

In fact, apply separation of variables to equation (45). Making the substitution

$$z = x + iy = re^{i\theta}$$

we reduce (45) to an equation in only two independent variables by intergrating both sides with respect to θ from 0 to 2π.

Consider the Fourier coefficient

$$V = ir\int_0^{2\pi} ue^{i\theta}d\theta = \oint_{|z|=r} u(x,y,t)dz \tag{46}$$

of any solution $u = u(x,y,t)$ of (45) as a function of the two independent variables: t and $\rho = r^2$.

Because of the divergence theorem we can have

$$V = i\iint_{|z|<r}\left(\frac{\partial u}{\partial x} + i\frac{\partial u}{\partial y}\right)dxdy\ .$$

Therefore from (46) we get

$$\begin{aligned}\frac{\partial V}{\partial r} &= i\oint_{|z|=r}\left(\frac{\partial u}{\partial x} + i\frac{\partial u}{\partial y}\right)ds \\ &= \int_{|z|=r}\left(\frac{\partial u}{\partial x} + i\frac{\partial u}{\partial y}\right)\frac{rdz}{z}\ .\end{aligned}$$

Thus from (45) we have

$$\frac{\partial V}{\partial \rho} = \frac{1}{2r}\frac{\partial V}{\partial r}$$
$$= \frac{1}{2} \cdot \oint_{|z|=r} \left(\frac{\partial u}{\partial x} + i\frac{\partial u}{\partial y} \right) \frac{dz}{z}$$
$$= i \oint_{|z|=r} \frac{\partial u}{\partial t} dz + \frac{1}{2} g'(t) \cdot \oint_{|z|=r} \frac{dz}{z}$$

Then from (46) we get

$$\frac{\partial V}{\partial t} + i\frac{\partial V}{\partial \rho} + \pi g'(t) = 0 \ .$$

This relation shows that expression

$$U = V + \pi g'(t) \tag{47}$$

satisfies Cauchy-Riemann equations

$$\frac{\partial U}{\partial t} + i\frac{\partial U}{\partial \rho} = 0 \ .$$

Therefore U is an *analytic* function of complex variable

$$\sigma = t + i\rho \ .$$

Then from (46) we have $V = 0$ when $\rho = 0$, so that

$$U(t) = \pi g(t) \tag{48}$$

on the real axis.

Because g is real-valued, (48) yields that the imaginary part of the analytic $U(\sigma)$ vanishes along the real axis. Hence by *Schwarz Reflection Principle* we get

$$U(\bar{\sigma}) = \overline{U(\sigma)} \tag{49}$$

for continuing U analytically across the line $\rho = 0$ into the lower halfplane.

Since U must be regular on the real axis, we conclude from (48) that $g = g(t)$ is analytic in some interval around $t = 0$. Thus $g' = g'(t)$ has to be analytic at $t = 0$.

Therefore equation (45) possesses *no* continuously differentiable solutions near the origin when $g'(t)$ is real, but *not* analytic, in an interval around $t = 0$.

Remarks

(i) We can associate choices of the term g' in (45) for which *no* continuously differentiable solutions of (45) exist in any neighborhood of (x_0, y_0, t_0): arbitrary point in space.

In particular, equation

$$\left[\left(\frac{\partial}{\partial x} + i\frac{\partial}{\partial y}\right) - 2i(x+iy)\frac{\partial}{\partial t}\right]u = g'(t - t_0 - 2y_0 x + 2x_0 y) \tag{50}$$

has *no* such solution u near (x_0, y_0, t_0) when the real function

$$g' = g'(t^*) ; \quad t^* = t - t_0 - 2y_0 x + 2x_0 y$$

fails to be analytic at $t^* = 0$.

In fact, consider variables t^* above and

$$x^* = x - x_0 \,, \; y^* = y - y_0 \,.$$

Then equation (50) is written

$$\left[\left(\frac{\partial}{\partial x^*} + i\frac{\partial}{\partial y^*}\right) - 2i(x^* + iy^*)\cdot\frac{\partial}{\partial t^*}\right]u = g'(t^*) \,. \tag{45$'$}$$

(ii) Consider *equation of H. Lewy*

$$\left[\left(\frac{\partial}{\partial x} + i\frac{\partial}{\partial y}\right) + (x - iy)\frac{\partial}{\partial t}\right]u = f \,. \tag{51}$$

This equation has *no* solutions for general $f \in \mathbf{C}^\infty$ (: the class of all infinitely differentiable complex functions on $\mathbb{R}^1$), even if we ask only for distribution solutions u defined in a small neighborhood.

4.15 *A Partial Differential Equation failing to be locally solvable.*

Consider Partial Differential Operator

$$L = \frac{\partial}{\partial t} + i \sum_{k=1}^{n} a_k(t) \frac{\partial}{\partial x_k} , \tag{52}$$

and equation

$$Lu = f . \tag{52)'$$

Equation (52)′ can be solved by making a partial Fourier transform in x-variables.

Therefore (52)′ goes over to an ordinary differential equation

$$\left[\frac{\partial}{\partial t} + \sum_k a_k(t) \xi_k \right] \hat{u}(t, \xi) = \hat{f}(t, \xi) . \tag{52)''$$

Equation (52)″ is solved by employing the integrating factor

$$\exp \left[\int^t \sum_k a_k(s) \xi_k ds \right] . \tag{53}$$

Thus there exists a formal solution $\hat{u} = \hat{u}(t, \xi)$ which grows exponentially, because integrating factor (53) grows exponentially in ξ.

In this case the partial Fourier transform u cannot be inverted, and (52)′ has *no* solutions (**19**, p. 138–139).

4.16 *The Initial Value Problem for the heat equation: $u_{xx} = u_t$ with initial condition: $u(x,0) = f(x)$ has no solution for $t < 0$ unless f is analytic.*

Note:

There exists a solution of equation

$$u_{xx} = u_t \tag{54}$$

of the form

$$u = u(x,t) = ae^{-\frac{t}{a^2}} \sin\left(\frac{x}{a^2}\right) \tag{55}$$

so that u approaches 0 for $t \geq 0$, but *not* for $t < 0$, in the limit as $a \to 0$ (**23**, p. 455–456).

4.17 *Domains (in three or more dimensions) for which an Elliptic boundary value problem has no solution in the strong sense.*

See: **14**, p. 303–305 and **42**, p. 370–372.

The following *Counter-example* is due to H. Lebesgue:

Claim that in three or more dimensions there exist domains for which the elliptic boundary value problem has *no* solution *in the strong sense*; that is, for prescribed countinuous boundary values we cannot always expect that these values are assumed at all the boundary points.

In fact, evaluate the *potential* of a mass distribution concentrated on the segment of x-axis between 0 and 1 with linear density $\tau(x) = x$, so that

$$\begin{aligned} u = u(x,y,z) &= \int_0^1 \frac{\xi d\xi}{\sqrt{(\xi - x)^2 + \rho^2}} \\ &= A(x,\rho) - 2x \ln \rho \, , \end{aligned} \tag{56}$$

where

$$\rho^2 = y^2 + z^2 \, ,$$

$$A(x,\rho) = \sqrt{(1-x)^2+\rho^2} - \sqrt{x^2+\rho^2}$$
$$+ x\ln\left|\left(1-x+\sqrt{(1-x)^2+\rho^2}\right)\left(x+\sqrt{x^2+\rho^2}\right)\right| .$$

Then $A = A(x,\rho) \to 1$, as the origin is approached through values of $x(>0)$.

However the limit of

$$B = B(x,\rho) = -2x\ln\rho$$

depends on the path of the approach.

Examples:

(**1**) Approach the origin on the surface

$$\rho = |x|^n .$$

Then $B = B(x,\rho) \to 0$ for every n, and hence

$$u \to 1 .$$

(**2**) Take the surface

$$\rho = \exp\left(-\frac{c}{2x}\right), \quad c>0\,, x>0\,,$$

which has an "infinitely sharp" peak at the origin.

Then $B = B(x,\rho) \to c$, and hence the *potential*

$$u \to 1+c .$$

Note: This means that all the equipotential surfaces

$$u = 1+c\,, \quad c>0$$

meet at the origin.

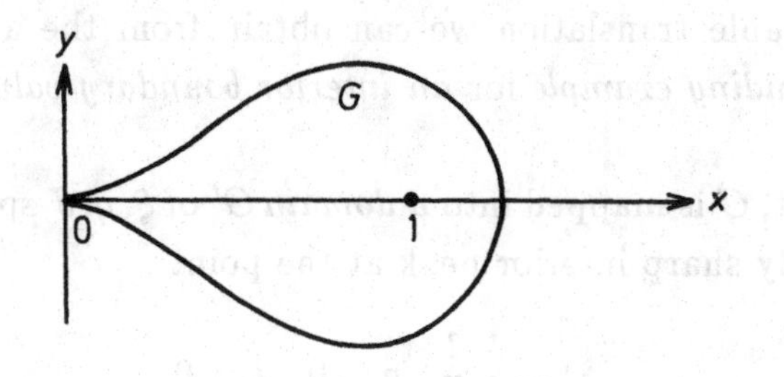

Fig. 4.4

All derivatives of the curve

$$\rho = f(x) \ ,$$

from which the above-mentioned surfaces are formed by rotation about the x-axis, vanish at the origin.

Such a surface

$$u = 1 + c \ , \quad c > 0 \tag{57}$$

is sketched in Figure 4.4.

Take *a fundamental domain* the region G: bounded by surface (57).

Solving *the exterior boundary value problem for "Laplace (or Potential) equation" (in G above)*

$$\Delta u = \left(\frac{\partial^2}{\partial x^2} + \frac{\partial^2}{\partial y^2} + \frac{\partial^2}{\partial z^2}\right) u = 0 \tag{58}$$

and boundary values (57), we find that formula (56) furnishes the regular solution of exterior problem (57)–(58).

Note: If we approach the origin in a suitable way, this solution u $\to$ (any value between 1 and 1 + c).

Second Counter-Example

By inversion with respect to the sphere

$$S : \left(x - \frac{1}{2}\right)^2 + y^2 + z^2 = \frac{1}{4}$$

and a suitable translation we can obtain from the above example *a corresponding example* for *an interior boundary value problem for* (58).

In fact, G is mapped into *a domain* G' of ξ, η, J-space which has an infinitely sharp interior peak at the point

$$\xi = -\frac{1}{2}\,, \eta = 0\,, J = 0\,.$$

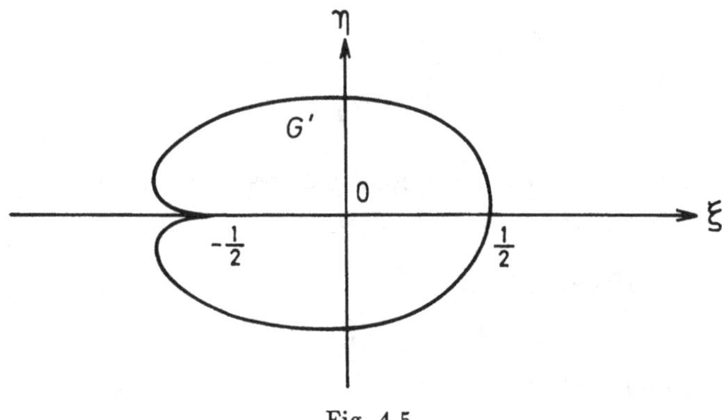

Fig. 4.5

The boundary values (57) go over into the boundary values

$$v = \frac{1+c}{2r}\,, \tag{59}$$

where

$$r = (\xi^2 + \eta^2 + J^2)^{\frac{1}{2}}\,,$$

continuous on the boundary curve $\Gamma' = \partial G'$.

Solving *the interior boundary value problem for (58)'* $: \Delta v = 0$ (in G' above) and boundary values (59), we find that formula

$$v = v(\xi, \eta, J) = \frac{1}{2r} u\left(\frac{\xi}{4r^2} + \frac{1}{2}\,, \frac{\eta}{4r^2}\,, \frac{J}{r^2}\right) \tag{60}$$

furnishes the regular solution of interior problem (58)$'$–(59).

Note: If we approach the point $\xi = -\frac{1}{2}, \eta = 0, J = 0$ in a suitable way, this solution $v \to$ (any value between 1 and 1 + c).

Remarks

(1) The above domains in 3-dimensional space are comparatively simple simply-connected domains irregular with respect to *the Dirichlet problem for Laplace equation.*

(2) *The Dirichlet problem* of finding a function harmonic in G (or in G') and regular in its closure $\overline{G}$ (or $\overline{G'}$), and satisfying the boundary condition (57) (or(59)) on the boundary $\Gamma = \partial G$ (or on $\Gamma' = \partial G'$), is *not* soluble in the class of functions regular in G (or in G').

Therefore the domain G (or G') is *not regular with respect to the Dirichlet problem*, because u(or v) is *not* continuous at the origin (or at the point: $\xi = -\frac{1}{2}, \eta = 0, J = 0$).

4.18 *A Dirichlet Problem for Laplace equation:* $\Delta u \equiv u_{xx} + u_{yy} = 0$ *in domain* $G: \ 0 < x^2 + y^2 < 1$ *and boundary values:*

$$u = \begin{cases} 0 \quad \text{on} \quad x^2 + y^2 = 1 \\ \\ 1 \quad \text{at} \quad (0,0) \end{cases} := f$$

without solution.

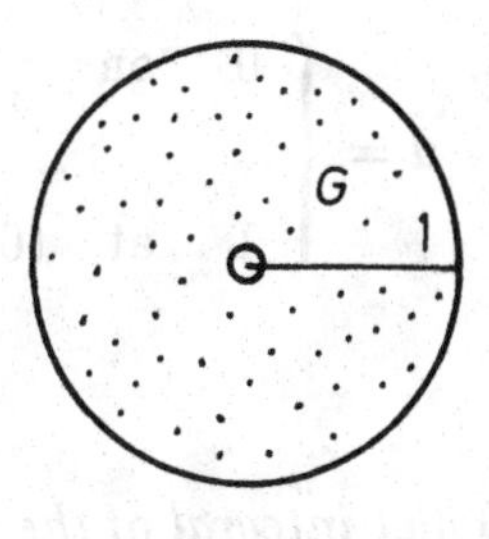

Fig. 4.6

In this case the condition that f ($u = f$ on $(x^2+y^2=1)\cup(0,0)$) is continuous is relaxed somewhat (**17**, p. 138–139).

By symmetry and uniqueness, it is obvious that if the Problem: $\Delta u = 0$, $u = f$ has a solution,it must have *a rotational symmetry* (that is, the solution must be a function of $r = \sqrt{x^2+y^2}$, independent of the polar angle θ).

It is clear that the solution of this Problem must be of the form

$$u = u(r) = a \ln r + b \ , \tag{60$'$}$$

where a, b: constants, and

$$u = \frac{1}{2} a \ln(x^2+y^2) + b \ ,$$

$$u_x = a\frac{x}{x^2+y^2} \ , \qquad u_{xx} = a\frac{-x^2+y^2}{(x^2+y^2)^2} \ ,$$

$$u_y = a\frac{y}{x^2+y^2} \ , \qquad u_{yy} = a\frac{x^2-y^2}{(x^2+y^2)^2} \ .$$

From condition $u = 0$ on $r = 1$ we get

$$b = 0 \ .$$

Consider $a = 0$. Then $u \equiv 0$.
If $a \neq 0$, then u is unbounded near (0,0).
Therefore in either case ($a = 0$, or $a \neq 0$) the 2nd condition of

$$u = \begin{cases} 0 & \text{on} \quad r = 1 \\ \\ 1 & \text{at} \quad (0,0) \end{cases}$$

is *not* satisfied.

4.19 *A divergent Dirichlet integral of the solution of a Dirichlet problem.*

The following is *Hadamard's Counter-example* (**17**, p. 198). Consider

$$u = u(r, \theta) = \sum_{n=1}^{\infty} \frac{r^{n!} \cos(n!\theta)}{n^2} \tag{61}$$

that is defined on $R : r^2 = x^2 + y^2 \leq 1$. This function $u = u(r, \theta)$ solves the Dirichlet problem for Laplace equation

$$\Delta u \equiv u_{xx} + u_{yy} = 0$$

and continuous boundary values

$$u = g(\theta) = \sum_{n=1}^{\infty} \frac{\cos(n!\theta)}{n^2} . \tag{62}$$

Then *the Dirichlet integral*

$$D(u) = \iint_R (u_x^2 + u_y^2) dx dy$$

diverges.

4.20 *A solution of $u_{xy} = 0$ in $D(= \mathbb{R}^2)$ is not solution of $u_{yx} = 0$ in D.*

A solution of equation: $u_{xy} = 0$ in the entire xy-plane is

$$u = f(x) \tag{63}$$

which is *not* solution of equation: $u_{yx} = 0$ in the above xy-plane, where $f = f(x)$ is (the first example of) a continuous everywhere nondifferentiable function given by K. W. T. Weierstrass (See: B. R. Gelbaum, and J. M. H. Olmsted,"Counter-examples in Analysis", Holden-Day,Inc., San Francisco, 1964 , p. 38–39)

$$f = f(x) = \sum_{n=0}^{\infty} b^n \cos(a^n \pi x) \tag{64}$$

where b is an odd integer, and a is such that $0 < a < 1$, and $ab > 1 + \frac{3}{2}\pi$.

Note (**9**, p. 26):

By a *solution* of a partial differential equation we mean a real-valued function in the class $\mathbf{C}^n(D)$, where

n is the order of the equation (that is, the highest of the orders of the partial derivatives that occur in the equation).

4.21 *A discontinuous function satisfying a partial differential equation without being a solution (in* $\mathbf{C}^1(\mathbb{R}^2)$*).*

Consider equation (**9**, p. 27)

$$x\frac{\partial u}{\partial x} + y\frac{\partial u}{\partial y} = 0 \tag{65}$$

in the entire xy-plane $D(= \mathbb{R}^2)$.

The function

$$u = u(x,y) = \begin{cases} \dfrac{xy}{x^2+y^2} & , \quad (x,y) \neq (0,0) \\ 0 & , \quad (x,y) = (0,0) \end{cases} \tag{66}$$

satisfies equation (65), and is discontinuous at (0,0) (therefore, it is *not* a solution in $\mathbf{C}^1(\mathbb{R}^2)$), because arbitrarily near (0,0) there exist points of the form (a,a) at which $u = \frac{1}{2} = u(\mathrm{a}, \mathrm{a})$.

See: B. R. Gelbaum, and J. M. H. Olmsted, "Counter-examples in Analysis", Holden–Day, Inc., San Francisco, 1964, p. 115.

Note:

The function

$$f = f(x,y) = \frac{xy}{x^2+y^2}, \quad (x,y) \neq (0,0) \tag{67}$$

above is *homogeneous of degree* 0 (that is, $f(tx,ty) = t^0 f(x,y)$ for all t).

Theorem 4.3

If $u = u(x, y)$ is a differentiable function of two variables and

$$u(tx, ty) = t^k u(x, y) \tag{68}$$

for all t (i.e., *u is homogeneous of degree k*),then it is clear that

$$x\frac{\partial u}{\partial x} + y\frac{\partial u}{\partial y} = ku\ . \tag{65$'$}$$

4.22 *"Never visit a universe where the temperature increases as* $\exp(x^2)$ *towards the boundary".*

Consider the boundary value problem (**9**, p. 27–28):

$$u_{xx} = u_t\ ,\ -\infty < x < \infty\ ,\ 0 < t < \delta \tag{69}$$

$$u(x, 0) = f(x)\ ,\ -\infty < x < \infty\ , \tag{70}$$

where $f = f(x)$ is in the class $\mathbf{C}(\mathbb{R})$.

A function $u = u(x, t)$ in the class $\mathbf{C}^2 = \mathbf{C}^2(-\infty < x < \infty\ ,\ 0 < t < \delta)$ is *a solution of* (69)–(70) if it satisfies (69) and if it has, in the band: $-\infty < x < \infty\ ,\ 0 \le t < \delta$, a continuous extension that for $t = 0$ coincides with $f(x)$.

A solution $u = u(x, t)$ of (69) shows, for each $x \in \mathbb{R} = (-\infty, \infty)$ and each t between 0 and δ, *the temperature at the time t at the point x of* an infinitely long homogeneous rod, if *the temperature distribution in the rod at the time $t = 0$ is* given by $f(x)$.

Claim that $\delta, f(x)$, and solution $u = u(x, t)$ can be given so that $u = u(x, t)$ *cannot be extended beyond $t = \delta$.*

In fact, if

$$f(x) = \exp(x^2)\ , \tag{71}$$

then the function

$$u = u(x, t) = (1 - 4t)^{-\frac{1}{2}} \exp\left(\frac{x^2}{1 - 4t}\right) \tag{72}$$

is a solution of Problem (69)–(70) for

$$\delta = \frac{1}{4} \tag{73}$$

but *not* for any larger value of δ.
Note: At time $t = \frac{1}{4}$ the considered universe becomes infinitely hot (provided that *no* relativistic effects are taken into account).

Second Counter-Example

In the above *Counter-example* (69)–(70) with choice (71) the solution (72) has the "bounded life-time" δ.

Now we can also choose

$$f(x) = \frac{1}{\sqrt{\pi}} \exp(-x^2) \tag{71$'$}$$

so that there exists a solution

$$u = u(x,t) = \frac{1}{\sqrt{\pi}}(1+4t)^{-\frac{1}{2}} \exp\left(-\frac{x^2}{1+4t}\right) \tag{72$'$}$$

with a "bounded past" δ for $t > -\frac{1}{4}$ but *not* for $t > t_0$ if $t_0 < -\frac{1}{4}$.

An Abstract Cauchy Problem

Consider equation (**18**, p. 29–30):

$$u'(t) = Au(t)\ , \quad t \geq 0\ , \tag{74}$$

where A is *densely* defined operator in *an arbitrary* (real or complex) *Banach space E* (that is, a normed linear space which is complete in the metric defined by its norm).

A *solution of* (74) is a function $t \to u(t)$ which is continuously differentiable for $t \geq 0$ and such that $u(t) \in D(A)$: Domain of A, satisfying equation (74) for $t \geq 0$ and initial (Cauchy) condition

$$u(0) = u_0\ . \tag{75}$$

The Problem (74)–(75) is *an Abstract Cauchy Problem.* An analogous Abstract Cauchy Problem we can have for $-\infty < t < \infty$.

4.23 *An Abstract Cauchy Problem for heat-diffusion equation without solution in finite time.*

Consider the heat-diffusion equation (**18**, p. 348):

$$\frac{\partial u}{\partial t} = k\frac{\partial^2 u}{\partial x^2} \ , \ -\infty < x < \infty \ , \ -\infty < t < T \ . \tag{74$'$}$$

Assume

$$E = \mathbf{C}_{a,2}(a > 0) = \{u = u(x) : -\infty < x < \infty \text{ , with norm}$$
$$\|u\| = \sup_{-\infty<x<\infty} |u(x)|(1+|x|)\exp(-ax^2) < \infty\} \ .$$

It is clear that each $\mathbf{C}_{a,2}$ is a Banach space.
Define the operator A by

$$Au = u'' \tag{76}$$

with domain

$$D(A) = \{u = u(x) : \text{twice differentiable in } E \text{ with } u'' \in E\} \ .$$

Let $0 < b < a$,

$$u(x,t) = \frac{1}{\sqrt{1-4bkt}} e^{\frac{bx^2}{1-4bkt}} \ . \tag{77}$$

Then the E-valued function

$$t \to u(\cdot\ , t)$$

is a solution, so called "*explosive solution*", of Cauchy Problem $(74)'$, $-\infty < t < T = \frac{a-b}{4abk}$ with initial condition

$$u(x,0) = e^{bx^2} \tag{78}$$

but *it ceases to exist for*

$$t = T$$

(that is, $\lim_{t \to T^-} \|u(t)\| = \infty$).

Second Counter-Example:

Consider

$E = \mathbf{BC}(-\infty, \infty) = \{u$: bounded continuous functions endowed with the supremum norm$\}$,

or

$E = \mathbf{L}^1(-\infty, \infty)$.

Then "explosive solutions" of $(74)'$ *do not exist.* In fact, in both cases ($\mathbf{BC}$, or $\mathbf{L}^1$) the solution corresponding to the initial condition $u(x)$ is given by *the Weierstrass formula*

$$u = u(x,t) = \frac{1}{2\sqrt{\pi k t}} \int_{-\infty}^{\infty} e^{-\frac{(x-\xi)^2}{4kt}} u(\xi) d\xi \ , \quad t > 0 \ . \tag{79}$$

4.24 *An Initial-Boundary Value Problem for the heat equation in multi-dimensional regions with nonlinear boundary data without global smooth solutions.*

Consider heat equation (**8**, p. 21–28):

$$\frac{\partial u}{\partial t} = \Delta u \left(= \frac{\partial^2 u}{\partial x_1^2} + \frac{\partial^2 u}{\partial x_2^2} + \ldots + \frac{\partial^2 u}{\partial x_n^2} \right) \tag{80}$$

in $\Omega \times [0, T)$, where $\Omega (\subseteq \mathbb{R}^n)$ is a bounded open domain with smooth boundary $\partial\Omega$, and $T > 0$.

Besides, assume nonlinear boundary data

$$\frac{\partial u}{\partial \nu} = f(u) \qquad \text{on} \qquad \partial\Omega \times [0, T) \tag{81}$$

$\nu = (\nu_1 , \nu_2 , \ldots , \nu_n)$ is the unit outward normal vector to $\partial\Omega$, $f \in \mathbf{C}^1(\mathbb{R}^1)$, and $\frac{\partial u}{\partial \nu} = \frac{\partial u}{\partial x_1}\nu_1 + \frac{\partial u}{\partial x_2}\nu_2 + \ldots + \frac{\partial u}{\partial x_n}\nu_n$ is the normal derivative of u on $\partial\Omega$.

Finally suppose initial data

$$u(x, 0) = u_0(x) \qquad \text{on} \qquad \bar{\Omega} , \tag{82}$$

where $u_0 = u_0(x) \in \mathbf{C}^2(\bar{\Omega})$.

Exhibit *a class of nonlinearities of* f *for which there does not exist a global solution* u *of initial-boundary value problem (80)–(82) on* $\bar{\Omega} \times [0, \infty)$. In fact, such an exhibition is given by

Theorem 4.4

Assume that

(i)

$$f = f(y) = |y|^{2a+1}h(y) \ , a > 0 \ ,$$

and $h = h(y)$ is monotone increasing (i.e.,$h'(y) > 0$)

(ii)

$$\oint_{\partial\Omega} \left(\int_0^{u_0(s)} f(y)dy \right) ds \geq \frac{1}{2} \int_{\Omega} |\nabla u_0|^2 dx \ .$$

Then there is *no* classical smooth solution $u : \bar{\Omega} \times [0, \infty) \to \mathbb{R}^1$ of Problem (80)–(82).

Proof

Suppose that there exists a classical smooth solution $u : \bar{\Omega} \times [0, T) \to \mathbb{R}^1$ of problem (80)–(82), for each $T > 0$.

Define

$$F = F(t) = \int_0^t \int_{\Omega} u^2(x, \eta)dxd\eta + (T - t) \cdot \int_{\Omega} u_0^2(x)dx + \beta(t + \tau)^2 \ ,$$

where $\beta \geq 0$, $\tau \geq 0$ are arbitrary real numbers.

Then

$$F'(t) = \int_\Omega u^2(x,t)dx - \int_\Omega u_0^2(x)dx + 2\beta(t+\tau)$$
$$= \int_0^t \left(\frac{\partial}{\partial\eta} \int_\Omega u^2(x,\eta)dx \right) d\eta + 2\beta(t+\tau)$$

or

$$F'(t) = 2\int_0^t \int_\Omega u\Delta u dx d\eta + 2\beta(t+\tau) . \tag{83}$$

But

$$\int_\Omega u\Delta u dx = \oint_{\partial\Omega} uf(u)ds - \int_\Omega |\nabla u|^2 dx$$

by Green's formula and the nonlinear boundary data (81). Therefore

$$F'(t) = -2\int_0^t \int_\Omega |\nabla u|^2 dx d\eta + 2\int_0^t \oint_{\partial\Omega} uf(u)ds d\eta + 2\beta(t+\tau) \tag{84}$$

and

$$F''(t) = -2\int_\Omega |\nabla u|^2 dx + 2\oint_{\partial\Omega} uf(u)ds + 2\beta . \tag{85}$$

But

$$\int_0^t \int_\Omega \nabla u_\eta \nabla u dx d\eta = \frac{1}{2}\int_0^t \frac{\partial}{\partial\eta}\int_\Omega \nabla u \nabla u dx d\eta$$
$$= \frac{1}{2}\int_\Omega |\nabla u|^2 dx - \frac{1}{2}\int_\Omega |\nabla u_0|^2 dx , \tag{86}$$

and

$$\int_0^t \int_\Omega \nabla u_\eta \nabla u dx d\eta$$
$$= \int_0^t \int_\Omega \frac{\partial u_\eta}{\partial x_i}\frac{\partial u}{\partial x_i} dx d\eta$$
$$= \int_0^t \int_\Omega \frac{\partial}{\partial x_i}\left(u_\eta \frac{\partial u}{\partial x_i}\right) dx d\eta - \int_0^t \int_\Omega u_\eta \Delta u dx d\eta$$
$$= \int_0^t \oint_{\partial\Omega} u_\eta f(u) ds d\eta - \int_0^t \int_\Omega u_\eta^2 dx d\eta . \tag{87}$$

Relations (86)–(87) yield

$$\int_\Omega |\nabla u|^2 dx = \int_\Omega |\nabla u_0|^2 dx + 2\int_0^t \oint_{\partial\Omega} u_\eta f(u) ds d\eta$$
$$-2\int_0^t \oint_\Omega u_\eta^2 dx d\eta \,. \tag{88}$$

Therefore we get from (85) and (88) that

$$F''(t) = -2\int_\Omega |\nabla u_0|^2 dx - 4\int_0^t \oint_{\partial\Omega} u_\eta f(u) ds d\eta$$
$$+4\int_0^t \int_\Omega u_\eta^2 dx d\eta + 2\oint_{\partial\Omega} u f(u) ds + 2\beta \,. \tag{89}$$

We split

$$2\beta = 4(\alpha+1)\beta - 2(2\alpha+1)\beta \,, \quad \alpha > 0 \tag{90}$$

and use (87) again. Then (89) takes the form

$$F''(t) = 4(\alpha+1)\left(\int_0^t \int_\Omega u_\eta^2 dx d\eta + \beta\right)$$
$$+2\left[-\int_\Omega |\nabla u_0|^2 dx - 2\int_0^t \oint_{\partial\Omega} u_\eta f(u) ds d\eta\right.$$
$$\left.-2\alpha\int_0^t \int_\Omega u_\eta \Delta u dx d\eta + \oint_{\partial\Omega} u f(u) ds - (2\alpha+1)\beta\right] \,. \tag{91}$$

Employing

$$\oint_{\partial\Omega} \int_0^t u_\eta f(u) d\eta ds = \oint_{\partial\Omega} \int_0^t \frac{\partial}{\partial\eta}\left(\int_{u_0(s)}^{u(s,\eta)} f(y) dy\right) d\eta ds \tag{92}$$

and Green's formula so that

$$-2\int_0^t \oint_{\partial\Omega} u_\eta f(u) ds d\eta - 2\alpha\int_0^t \int_\Omega u_\eta \Delta u dx d\eta$$
$$= -2(\alpha+1)\int_0^t \oint_{\partial\Omega} u_\eta f(u) ds d\eta$$
$$+\alpha\left[\int_\Omega |\nabla u|^2 dx - \int_\Omega |\nabla u_0|^2 dx\right] \tag{93}$$

formula (91) is written as follows

$$\begin{aligned} F''(t) &= 4(\alpha+1)\Big(\int_0^t \int_\Omega u_\eta^2 dx d\eta + \beta\Big) \\ &+ 2\Big\{\alpha \int_\Omega |\nabla u|^2 dx - (\alpha+1)\int_\Omega |\nabla u_0|^2 dx + \oint_{\partial\Omega} \Big[u f(u) \\ &- 2(\alpha+1)\int_0^t \Big(\frac{\partial}{\partial\eta}\int_{u_0(s)}^{u(s,\eta)} f(y)dy\Big)d\eta\Big] ds \\ &- (2\alpha+1)\beta\Big\} = F''_{(1)}(t) + F''_{(2)}(t) \,, \end{aligned} \tag{94}$$

where

$$F''_{(1)}(t) = 4(\alpha+1)\left(\int_0^t \int_\Omega u_\eta^2 dx d\eta + \beta\right) \,, \tag{95}$$

and

$$F''_{(2)}(t) = F''(t) - F''_{(1)}(t) \,. \tag{96}$$

Consider

$$\hat{F}(t) = \int_0^t \int_\Omega u^2 dx d\eta + \beta(t+\tau)^2 < F(t) \,, \quad t < T \,. \tag{97}$$

Then

$$\begin{aligned} FF'' - (\alpha+1)(F')^2 &= F(F''_{(1)} + F''_{(2)}) - (\alpha+1)(F')^2 \\ &\geq (\hat{F}F''_{(1)} - (\alpha+1)(F')^2) + FF''_{(2)} \,. \end{aligned} \tag{98}$$

Consider

$$\begin{aligned} S^2 &= \left(\int_0^t \int_\Omega u^2 dx d\eta + \beta(t+\tau)^2\right)\left(\int_0^t \int_\Omega u_\eta^2 dx d\eta + \beta\right) \\ &- \left(\int_0^t \int_\Omega u u_\eta dx d\eta + \beta(t+\tau)\right)^2 \\ &\geq 0 \quad \text{(by Cauchy-Schwarz inequality)} \,, \end{aligned} \tag{99}$$

so that

$$\hat{F}F''_{(1)} - (\alpha+1)(F')^2 = 4(\alpha+1)S^2 \ . \tag{100}$$

Thus from (98) and (100) we get

$$FF'' - (\alpha+1)(F')^2 \geq FF_{(2)} \ . \tag{101}$$

Employing

$$\oint_{\partial\Omega} uf(u)ds = \oint_{\partial\Omega}\int_0^u \frac{d}{dy}(yf(y))dyds \tag{102}$$

and

$$\begin{aligned} &-2(\alpha+1)\oint_{\partial\Omega}\int_0^t\left(\frac{\partial}{\partial\eta}\int_{u_0(s)}^{u(s,\eta)} f(y)dy\right)d\eta ds \\ &= 2(\alpha+1)\oint_{\partial\Omega}\int_0^{u_0(s)} f(y)dyds \\ &- 2(\alpha+1)\oint_{\partial\Omega}\int_0^{u(s,t)} f(y)dyds \ , \end{aligned} \tag{103}$$

we get from (101) that

$$\begin{aligned} &FF'' - (\alpha+1)(F')^2 \\ &\geq 2F\left\{\alpha\int_\Omega|\nabla u|^2dx - (2\alpha+1)\beta + 2(\alpha+1)\left[\oint_{\partial\Omega}\int_0^{u_0(s)} f(y)dyds\right.\right. \\ &\left.\left.-\frac{1}{2}\int_\Omega|\nabla u_0|^2dx\right] + \oint_{\partial\Omega}\int_0^{u_0(s,t)}[yf'(y)-(2\alpha+1)f(y)]dyds\right\} . \end{aligned} \tag{104}$$

From our assumption

$$f(y) = |y|^{2\alpha+1}h(y) \ , \quad h'(y) > 0 \ ,$$

we get

$$\oint_{\partial\Omega}\int_0^u [yf'(y)-(2\alpha+1)f(y)]dyds > 0 \ . \tag{105}$$

Therefore (104) takes the form

$$FF'' - (\alpha+1)(F')^2 \geq 2F\left\{ -(2\alpha+1)\beta + 2(\alpha+1)\left[\oint_{\partial\Omega} \int_0^{u_0(s)} f(y)dyds - \frac{1}{2}\int_\Omega |\nabla u_0|^2 dx \right]\right\} . \tag{106}$$

Choose

$$\beta = \frac{2(\alpha+1)}{2\alpha+1}\left[\oint_{\partial\Omega}\left(\int_0^{u_0(s)} f(y)dy\right) ds - \frac{1}{2}\int_\Omega |\nabla u_0|^2 dx\right] \tag{107}$$

≥ 0 (From hypothesis (ii)).

Then

$$FF'' - (\alpha+1)(F')^2 \geq 0 \tag{108}$$

or

$$(F^{-\alpha})'' \leq 0 , \quad \alpha > 0 , \tag{108$'$}$$

which implies

$$F^\alpha(t) \geq F^{\alpha+1}(0)\Big[F(0) - \alpha t F'(o)\Big]^{-1} , \tag{109}$$

where

$$F'(0) = 2\beta\tau$$

and

$$F(0) = T\int_\Omega u_0^2(x) + \beta\tau^2 .$$

Thus from (109) we find

$$\lim_{t\to t_\infty} F(t) = +\infty , \tag{110}$$

where

$$t_\infty = \frac{T \int_\Omega u_0^2 dx + \beta\tau^2}{2\alpha\beta\tau} < T \, . \tag{111}$$

Relation (111) implies

$$\beta\tau^2 < T \left[2\alpha\beta\tau - \int_\Omega u_0^2(x) dx \right] \tag{112}$$

for sufficiently large T, provided we choose constant

$$\tau > \int_\Omega \frac{1}{2\alpha\beta} u_0^2(x) dx \ ,$$

completing the proof of Theorem.

5. UNIQUENESS

5.1 *An Initial Value Problem with at least two solutions*

Consider the ordinary differential equation

$$\frac{dy}{dx} = \sqrt{|y|} \tag{1}$$

with initial condition

$$y(0) = (0)\ . \tag{2}$$

The Initial Value Problem (1)–(2) has two solutions. In fact, the first solution is the x-axis (that is $y = 0$), but there is another one which is (**31**, p. 13; **30**, p. 66; **75**, p. 144):

$$y = \begin{cases} \dfrac{1}{4}x^2 & , \quad x \geq 0 \\ -\dfrac{1}{4}x^2 & , \quad x \leq 0\ . \end{cases}$$

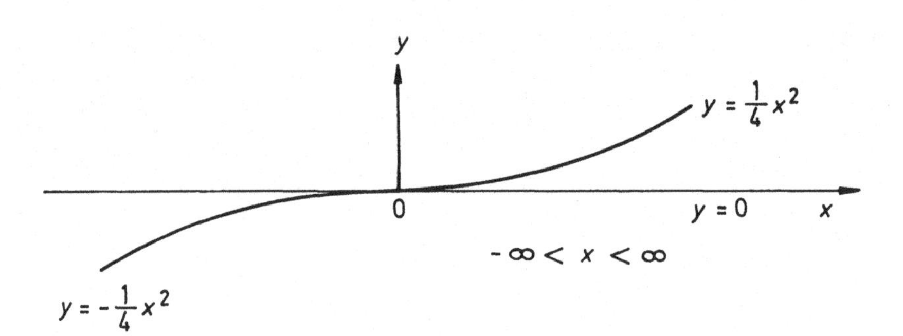

Fig. 5.1

Note: Equation (1) is so that the Lipschitz condition is *not* satisfied in any region which contains the x-axis, and admits of two real continuous solutions satisfying (2).

If equation (1) were given with $y \geq 0$; that is,

$$\frac{dy}{dx} = \sqrt{y} \,, \tag{1$'$}$$

then the Initial Value Problem (1)′–(2) would have the two solutions:

$$y = 0 \,,$$
$$y = \frac{1}{4}x^2 \,.$$

Correct the *error* of equation $y' = y^{1/2}$ (See: **31**, p. 13) so that it becomes $y' = |y|^{1/2}$.

Second Counter-Example

Consider equation (**78**, p. 34):

$$\frac{dy}{dx} = x\sqrt{y} \quad . \tag{3}$$

with initial condition (2), and $y \geq 0$.

The Initial Value Problem (3)–(2) has two solutions. In fact, the first solution is the x-axis, but there is a second one,

$$y = \frac{1}{16}x^4 \,.$$

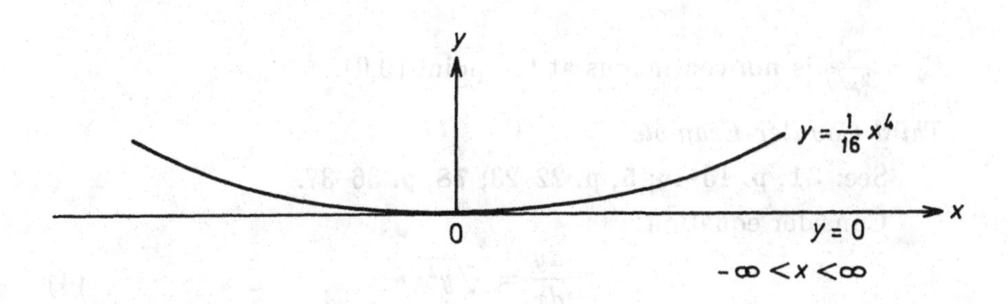

Fig. 5.2

Remarks:

(**i**) The Initial Value Problem (1)–(2) has infinitely many distinct solutions, for if a and b are arbitrary positive numbers, the function

$$y = \begin{cases} 0 & , \quad -b \le x \le a \\ \frac{1}{4}(x-a)^2 & , \quad x \ge a \\ -\frac{1}{4}(x+b)^2 & , \quad x \le -b \end{cases}$$

is everywhere continuous, differentiable, and a solution of (1)–(2) (See: **75**, p. 144).

(**ii**) The Initial Initial Value Problem (1)′–(2) has infinitely many distinct solutions, for if c is arbitrary positive number, the function

$$y = \begin{cases} 0 & , \quad x \le c \\ \frac{1}{4}(x-c)^2 & , \quad c < x \end{cases}$$

is everywhere continuous, differentiable, and a solution of (1)′–(2) (See: **20**, p. 13).

Observe that, if

$$F = F(x,y) = \sqrt{y} \, ,$$

$F_y = \frac{1}{2\sqrt{y}}$ is *not* continuous at the point (0,0).

Third Counter-Example

See: **31**, p. 15–16; **5**, p. 22–23; **78**, p. 36–37.

Consider equation

$$\frac{dy}{dx} = \sqrt[3]{y^2} \tag{4}$$

with initial condition (2).

The Initial Value Problem (4)–(2) has infinitely many distinct solutions, for if a and b are arbitrary positive numbers, the function

$$y = \begin{cases} 0 & , \quad a \le x \le b \\ \frac{1}{27}(x-b)^3 & , \quad x > b \\ (x-a)^3 & , \quad x < a \end{cases}$$

is everywhere continuous, differentiable, and a solution of (4)–(2) for any $a < b$. In addition, the null function

$$y = 0$$

also satisfies (4)–(2). Its graph is the *envelope* of the curves

$$y = \frac{1}{27}(x-c)^3 ,$$

where c is arbitrary positive number.

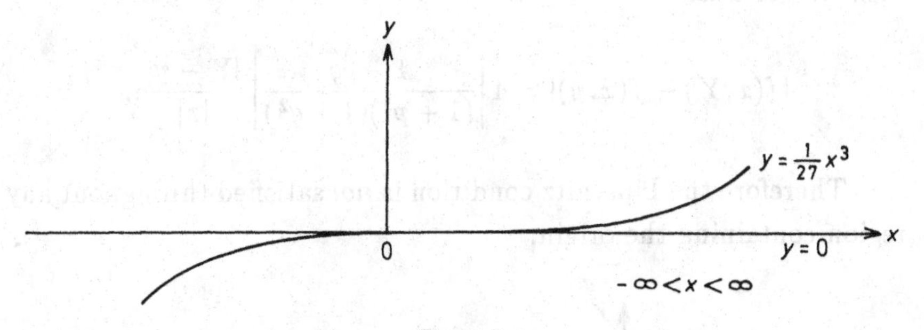

Fig. 5.3

Fourth Counter-Example

See: **31**, p. 13–14; **30**, p. 67.

Consider equation

$$\frac{dy}{dx} = \begin{cases} \frac{4x^3 y}{x^4 + y^2} & , \quad x \ne 0 , \; y \ne 0 , \\ 0 & , \quad x = 0 , \; y = 0 \end{cases} \tag{5}$$

with initial condition (2).

The Initial Value Problem (5)–(2) has inifinitely many distinct solutions, for if c is arbitrary positive number, the function

$$y = c^2 - \sqrt{x^4 + c^2}$$

is a solution of (5)–(2).

It is easily proved that $f = f(x, y)(= \frac{dy}{dx}$ given by (5)) is a continuous function of x and y. On the other hand,

$$f(x, Y) - f(x, y) = \frac{4x^3(x^4 - yY)}{(x^4 + y^2)(x^4 + Y^2)}(Y - y) .$$

If

$$y = px^2 , \quad Y = qx^2 ,$$

then we get that

$$|f(x, Y) - f(x, y)| = 4 \left| \frac{1 - pq}{(1 + p^2)(1 + q^2)} \right| \frac{|Y - y|}{|x|} .$$

Therefore the Lipschitz condition is *not* satisfied throughout any region containing the origin.

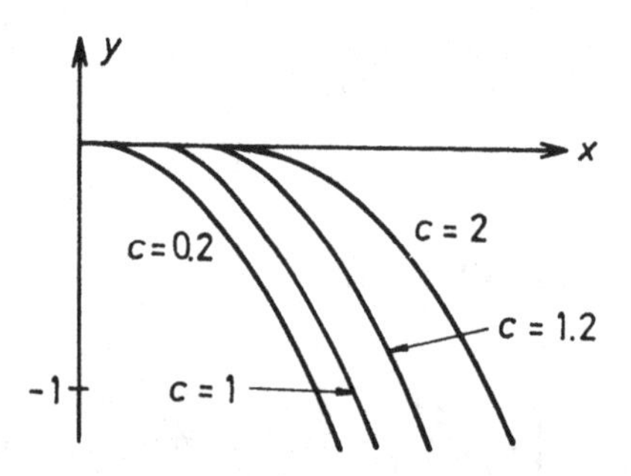

Fig. 5.4

Remarks:

(i) Concerning equation (4), the function

$$f(x,y) = \sqrt[3]{y^2}\left(= \frac{dy}{dx}\right)$$

satisfies a Lipschitz condition in any strip

$$D : |y| \geq \epsilon \ , \quad \epsilon > 0 \ ,$$

with *Lipschitz constant*

$$L = \sup_D \left| \frac{\partial f}{\partial y} \right|$$

$$= \frac{2}{3}\epsilon^{-1/3} \quad \text{(See: } \mathbf{5}\text{, p. 23)} \ .$$

(ii) The Initial Value Problem for equation (4) and initial condition

$$y(0) = -1 \tag{2$'$}$$

has *infinitely many distinct solutions*, for if a is arbitrary positive number, and $c > 3$, the function

$$y = \begin{cases} 0 & , \quad 3 \leq x \leq a \\ \dfrac{1}{27}(x-a)^3 & , \quad a < x \leq c \\ \dfrac{1}{27}(x-3)^3 & , \quad 0 \leq x \leq 3 \end{cases}$$

is a solution of (4)–(2)$'$.

(iii) Uniqueness breaks down for(4)–(2) because

$$f(x,y) = \sqrt[3]{y^2}$$

does *not* satisfy a Lipschitz condition in the neighborhood of the origin (See: **31**, p. 17–18).

In fact, take the points $(x, 0)$, (x, Y). Then

$$|f(x,Y) - f(x,0)| = |Y|^{-1/3}|Y - 0| .$$

By choosing Y sufficiently small it is clear that we can make $|Y|^{-1/3}$ larger than any preassigned constant.

Therefore Lipschitz condition fails to hold.

5.2 *The Differential Equation:* $(\frac{dx}{dt})^2 - 2\frac{dx}{dt} + 4x = 4t - 1$ *with initial conditions:* $x(0) = 0$, $x'(0) = 1$ *with two solutions.*

Consider (**12**, p. 40).

Theorem: Let $F = F(t, x, y)$ be a real continuous function of (t, x, y) in a real domain D containing the point t_0 , x_0 , $y_0)$. Let $\frac{\partial F}{\partial x}$ and $\frac{\partial F}{\partial y}$ exist and be continuous in D, and suppose

$$F(t_0 \, , x_0 \, , y_0) = 0 \, , \quad \frac{\partial F}{\partial y}(t_0 \, , x_0 \, , y_0) \neq 0 \, .$$

Then there exists a *unique* function

$$\varphi = \varphi(t)$$

being *a solution of* $F(t, x, y) = 0$ on some interval containing t_0, satisfying

$$F(t \, , \varphi(t) \, , \varphi'(t)) = 0 \, ,$$

and

$$\varphi(t_0) = x_0 \, , \quad \varphi'(t_0) = y_0 \, .$$

Note: The above theorem may fail where

$$F(t, x, y) = 0 \text{ ,and } \quad \frac{\partial F}{\partial y}(t, x, y) = 0 \, . \tag{6}$$

A solution of

$$F(t,x,x') = 0$$

may satisfy both equations (6) but *uniqueness may fail.*

Counter-Example

Consider equation

$$\left(\frac{dx}{dt}\right)^2 - 2\frac{dx}{dt} + 4x = 4t - 1 \tag{7}$$

with initial conditions

$$x(0) = 0\ , \quad x'(0) = 1\ . \tag{8}$$

It is clear that the Initial Value problem (7)–(8) has two solutions:

$$\begin{aligned} x &= x(t) = t\ , \\ x &= x(t) = t - t^2\ . \end{aligned}$$

Observe that

$$\begin{aligned} F = F(t,x,y) &= y^2 - 2y + 4x - (4t-1) = 0\ , \\ \frac{\partial F}{\partial y} &= \quad 2y - 2\ . \end{aligned}$$

Therefore

$$F(0,0,1) = 0\ , \text{and} \quad \frac{\partial F}{\partial y}(0,0,1) = 0\ .$$

5.3 *An Initial Value Problem: $x' = f(t,x)$, $x(\tau) = \xi$ with convergence of successive approximations insufficient for the uniqueness of solution.*

Consider (**12**, p. 3, 19, 45, 53–54 and **29**, p. 40–41).

Theorem: Let $f \in C$ on a domain D in (t,x) space be defined on the rectangle

$$R : |t-\tau| \leq a \ , \ |x-\xi| \leq b \quad (a,b>0)$$

about the point (τ,ξ). Given any $\epsilon > 0$, there exists *an ϵ - approximate solution* $\varphi = \varphi(t)$ of the Initial Value Problem

$$\frac{dx}{dt} = f(t,x) \ , \tag{9}$$

$$x(\tau) = \xi \tag{10}$$

on $|t-\tau| \leq a$, where

Definition: ϵ - *approximate solution* of equation (9) on a t interval I is a function $\varphi = \varphi(t) \in C$ on I such that

(i) $(t,\varphi(t)) \in D \quad (t \in I)$,

(ii) $\phi \in C^1$ on I, except possibly for a finite set of points S on I where φ' may have simple discontinuities (i.e., if the right and left limits exist but are *not* equal),

(iii) $|\varphi'(t) - f(t,\varphi(t))| \leq \epsilon \qquad (t \in I - S)$.

Successive Approximations:

For the initial point (τ,ξ) and equation (9) *the successive approximations* $\{\varphi_m(t)\}_{m\geq 0}$ are defined by

$$\varphi_0(t) = \xi \ , \tag{11}$$

$$\varphi_{m+1}(t) = \xi + \int_\tau^t f(s,\varphi_m(s))ds \ , \tag{12}$$

$$m = 0,1,2,\dots \ .$$

Consider

$$f = f(t,x) = \sqrt[3]{x} \ . \tag{13}$$

For initial point $(\tau, \xi) = (0,0)$ the successive approximations for equation (9) and (13) are all the zero functions such that

$$\varphi_0(t) = 0 ,$$

$$\varphi_{m+1}(t) = \int_0^t \sqrt[3]{\varphi_m(s)}ds = 0 , \quad m = 0, 1, 2, \ldots ,$$

and hence they converge to the zero function (solution).

On the other hand, the function $\varphi = \varphi(t)$ defined by

$$\varphi(t) = (\frac{2}{3}t)^{\frac{3}{2}}$$

is *another solution* which exists to the right of the origin.

Remarks:

(i) *The continuity of $f = f(t,x)$ in (9) is insufficient for the convergence of the successive approximations.*

In fact, let f be defined by

$$f = \begin{cases} 0 & , \quad t = 0 , \quad -\infty < x < \infty , \\ 2t & , \quad 0 < t \leq 1 , \quad -\infty < x < 0 , \\ 2t - \dfrac{4x}{t} & , \quad 0 < t \leq 1 , \quad 0 \leq x \leq t^2 , \\ -2t & , \quad 0 < t \leq 1 , \quad t^2 < x < +\infty \end{cases}$$

on the region : $0 \leq t \leq 1 , \quad -\infty < x < \infty$.

It is clear that $f = f(t,x)$ above mentioned is continuous and bounded by the constant 2.

For the initial point $(\tau, \xi) = (0,0)$ and equation (9) the successive approximations are

$$\varphi_0(t) = 0$$

$$\varphi_{2m-1} = t^2 , \ \varphi_{2m}(t) = -t^2 , \quad m = 1, 2, \ldots$$

for $0 \leq t \leq 1$.

Therefore the sequence $\{\varphi_m(t)\}$ has *two* cluster values for each $t \neq 0$, and hence the above successive approximations do *not* converge.

Note: Neither of two convergent subsequences $\{\varphi_{2m-1}(t)\}$, $\{\varphi_{2m}(t)\}$ converges to a solution, for

$$\varphi'_{2m-1}(t) = 2t \neq f(t, t^2), \text{ and } \varphi'_{2m}(t) = -2t \neq f(t, -t^2) .$$

(ii) Consider

$$f = f(t,x) = |x|^{-\frac{3}{4}} x + t \sin\left(\frac{\pi}{t}\right) ,$$

with $x(0) = 0$. Claim that :

If polygonal approximate solutions are set up (as in above Theorem) they need not converge as $\epsilon \to 0$.

In fact, let $t \geq 0$ and

$$t_k = k\delta \; , \quad k = 0, 1, 2, \dots ,$$

where

$$\delta = (n + \frac{1}{2})^{-1}$$

for some large n.

If n is *even*, we show that *the polygonal solution* $\varphi_n(t)$ satisfies

$$\varphi_n(\delta) = 0 ,$$
$$\varphi_n(2\delta) = \delta^2 , \quad \varphi_n(3\delta) > \frac{1}{2}\delta^{\frac{3}{2}} .$$

Once $\varphi_n(t) \geq \frac{1}{6} t^{\frac{3}{2}}$, it stays there as long as $t < \frac{1}{2000}$. Indeed, for $t \geq 4\delta$ and as long as

$$\varphi_n(t) \geq \frac{1}{6} t^{\frac{3}{2}} ,$$

$$\varphi_n'(t) > \varphi_n^{\frac{1}{4}}(t-\delta) - t > \frac{1}{2}(t-\delta)^{\frac{3}{8}} - t$$
$$> \frac{1}{10}t^{\frac{3}{8}} > \frac{d}{dt}\left(\frac{1}{6}t^{\frac{3}{2}}\right)$$

and the result follows.

If n is *odd*,

$$\varphi_n(t) < -\frac{1}{6}t^{\frac{3}{2}}$$

for $3\delta < t < \dfrac{1}{2000}$.

5.4 *A region in the xy-plane in which the Initial Value Problem:* $\dfrac{dy}{dx} = 3x\sqrt[3]{y}$, $y(0) = 0$ *has more than one solution.*

See: **64**, p. 49–50.

The function

$$f = f(x,y) = 3x\sqrt[3]{y} \tag{14}$$

is continuous in the entire xy-plane. However its partial derivative with respect to y

$$\frac{\partial f}{\partial y} = \frac{x}{\sqrt[3]{y^2}} \tag{15}$$

fails to exist along the line $y = 0$.

Theorem:

Let $f = f(x,y)$ and its partial derivative with respect to y be continuous throughout some rectangular region of the xy-plane containing the point (x_0, y_0). Then on some interval $x_0 - h < x < x_0 + h$ there is one and only one solution for the Initial Value Problem

$$\frac{dy}{dx} = f(x,y)\,, \quad y(x_0) = y_0\,. \tag{16}$$

Note that function (14) satisfies the hypotheses of the theorem in *any* rectangle that does *not* contain any part of the x-axis.

Any initial condition

$$y(0) = 0 \tag{17}$$

can be satisfied by *many solutions* of

$$\frac{dy}{dx} = 3x\sqrt[3]{y} \,. \tag{18}$$

For instance, both

$$y = 0 \,, \quad y = x^3 \tag{19}$$

satisfy the differential equation (18) and the initial condition (17).

Therefore the required region is *any rectangle that does not contain any part of the x-axis.*

5.5 *An Initial Value Problem related to motion of an object dropped in vacuum from an altitude x_0, with two solutions.*

Consider (**76**, p. 25,28)

Theorem:

Let $f = f(t,x)$ and $\frac{\partial f}{\partial x}$ be continuous on a region D in the tx-plane and let $(t_0 \,, x_0)$ be given interior point of D.

There is one and only one solution

$$x = \varphi(t) \,, \quad \tau_\varphi^- < t < \tau_\varphi^+ \,,$$

of the differential equation

$$\frac{dx}{dt} = f(t,x) \tag{20}$$

with the following properties:

(i) $\varphi(t_0) = x_0$,

(ii) $(t, \varphi(t)) \in D$, $\tau_\varphi^- < t < \tau_\varphi^+$,

(iii) either $|t| + |\varphi(t)| \to +\infty$, or $(t, \varphi(t))$ approaches a boundary point of D as $t \to \tau_\varphi^-$ and as $t \to \tau_\varphi^+$.

Note that the altitude $x = x(t)$ of the object satisfies the Initial Value Problem

$$\frac{dx}{dt} = -\sqrt{2g(x_0 - x)} \ , \tag{21}$$

$$x(0) = x_0 \ . \tag{22}$$

One solution of this Problem (21)–(22) is the constant solution

$$x = x_0 \ . \tag{23}$$

This is *not* the solution that describes the descent of the object. *The physically meaningful solution* is given by the function

$$x = x_0 - \frac{1}{2} g t^2 \ . \tag{24}$$

where g (= constant) is the acceleration due to gravity. In fact,

$$\lim_{\epsilon \to x_0} \left(\int_\epsilon^x \frac{du}{\sqrt{2g(x_0 - u)}} \right) = -\int_0^t ds \ , \quad \text{or}$$

$$\sqrt{\frac{2(x_0 - x)}{g}} = t \ , \quad \text{or}$$

(24) holds.

Therefore the Initial Value Problem (21)–(22) has *two solutions* (23)–(24).

Theorem above is *not* contradicted, however, since

$$\frac{\partial}{\partial x}(\sqrt{2g(x_0 - x)}) = -\frac{g}{\sqrt{2g(x_0 - x)}}$$

is *discontinuous* at $x = x_0$.

5.6 *A Function $U = U(t,u)$ Continuous on the (t,u)-plane such that the Initial Value Problem: $u' = U(t,u)$, $u(t_0) = u_0$ has more than one solution on every interval $[t_0, t_0+\epsilon]$ and $[t_0-\epsilon, t_0]$ for arbitrary $\epsilon > 0$ and for every choice of initial point (t_0, u_0).*

It is due to Ph. Hartman (**29**, p. 15–23).

Consider the following Theorem due to H. Kneser, which concerns *nonunique solutions.*

Theorem:

Let $f = f(t,y)$ be continuous on

$$R : t_0 \leq t \leq t_0 + a \; , \quad |y - y_0| \leq b \; .$$

and let $|f(t,y)| \leq M$, $\alpha = \min(a, \frac{b}{M})$ and $t_0 < c \leq t_0 + \alpha$.

Finally, let S_c be the set of points y_c for which there is a solution $y = y(t)$ of

$$y' = f(t,y) \; , \tag{25}$$

$$y(t_0) = y_0 \tag{26}$$

on $[t_0, c]$ such that $y(c) = y_c$.

Then S_c is a *continuum* (that is, a closed and connected set).

The Proof of this Theorem is omitted.

Ph. Hartman found the required function $U = U(t,u)$ as follows:

Let S_0 be the set of arcs:

$$u = 4i + \cos(\pi t) \quad \text{and} \quad u = 4i + 2 - \cos(\pi t) \tag{27}$$

for $-\infty < t < \infty, i = 0, \pm 1, \ldots$ considered to be made up of subarcs defined on the intervals of length 1, $k \leq t \leq k+1$ and $k = 0, \pm 1, \ldots$.

For every $n = 0, 1, 2, \ldots$, there will be constructed a set S_n of twice continuously differentiable arcs

$$u = u_{jk}(t) \; , \quad \frac{k}{2^n} \leq t \leq \frac{k+1}{2^n} \; , \text{and} \quad j,k = 0, \pm 1 \; , \ldots \; . \tag{28}$$

The symbol S_n will denote either the set of arcs (28) or the set of the points on these arcs.

The set S_n of arcs (28) will have the *properties* that

(i) $u_{jk}(t) < u_{j+1,k}(t)$ for $\frac{k}{2^n} < t < \frac{k+1}{2^n}$;

(ii) The arcs $u = u_{jk}(t)$ and $u = u_{j+1,k}(t)$ have exactly one end-point in common;

(iii) For any pair (j,k), there is at least one index h such that $u_{h,k-1} = u_{h+1,k-1} = u_{jk}$ at $t = \frac{k}{2^n}$ and an index i such that

$$u_{i,k+1} = u_{i+1,k+1} = u_{jk} \quad \text{at} \quad t = \frac{k+1}{2^n} \ ;$$

(iv) Any two arcs of S_n which have a point in common have the same tangent at that point; hence

(v) Any continuous arc $u = u(t)$, say, on $a \leq t \leq b$,which is made up of arcs of S_n can be continued over $-\infty < t < \infty$,*not uniquely*, so as to have the same property and any such continuation is of class $\mathbf{C}^1$ (and piecewise of class $\mathbf{C}^2$); also,

(vi) If $U_n = U_n(t,u)$ is defined on the point set S_n to be the slope of the tangent at the point $(t,u) \in S_n$, then U_n is uniformly continuous on S_n and the arcs of (v) constitute the set of the solutions of

$$u' = U_n(t,u) \ ; \tag{29}$$

(vii) The sets $S_0, S_1, \ldots$ satisfy $S_n \subset S_{n+1}$, so that $U_{n+1} = U_{n+1}(t,u)$ is an extension of U_n;

(viii): $S = \cup S_n$ and, in fact, the set of the end points $(\frac{k}{2^n}, u_{jk}(\frac{k}{2^n}))$ for $j,k = 0, \pm 1 \ldots$ and $n = 0,1,\ldots$, is dense in the plane; finally,

(ix)

$$U = U(t,u) = \lim_{n\to\infty} U_n \tag{30}$$

which is defined on S has a (unique) extension over the plane. Condition (ix) is the only nontrivial solution.

Let $\pi^2 = \epsilon_0 > \epsilon_1 > \dots$,

$$M_n = \sum_{k=0}^{n} \epsilon_k \qquad \text{and} \qquad M = \sum_{k=0}^{\infty} \epsilon_k (< \infty) \, . \tag{31}$$

Suppose that S_n has already been constructed so that the functions (28) satisfy

$$\left|u'_{jk}(t)\right| , \left|u''_{jk}(t)\right| \leq M_n \; ; \tag{32}$$

and that if

$$d_n = \sup_{j,k,t} \max \left(\left|u_{j+1,k}(t) - u_{j,k}(t)\right| , \left|u'_{j+1,k}(t) - u'_{jk}(t)\right| \right) ,$$

then

$$d_n \leq \epsilon_n \; ; \tag{33}$$

and, if $n > 0$ and no arc of S_{n-1} lies between $u = u_{ik}(t)$, $u = u_{hk}(t)$, then

$$|u'_{ik}(t) - u'_{hk}(t)| \leq d_{n-1} + \epsilon_n \, . \tag{34}$$

The set of the arcs S_{n+1} will be obtained from those S_n by inserting on each interval,

$$\left[\frac{k}{2^n} , \frac{2k+1}{2^{n+1}}\right] \quad \text{and} \quad \left[\frac{2k+1}{2^{n+1}} , \frac{k+1}{2^n}\right] ,$$

a finite number of arcs between the arcs

$$u = u_{jk}(t) \, , \; u = u_{j+1,k}(t) \quad \text{of} \quad S_n \, .$$

The arcs of S_n and these inserted arcs will constitute the set S_{n+1}.

For convenience, let

$$u = u(t) = u_{jk}(t) \, , \; v = v(t) = u_{j+1,k}(t) \, ,$$

$$a = \frac{k}{2^n} , \ b = \frac{k+1}{2^n} , \ c = \frac{a+b}{2} .$$

Suppose $u(a) = v(a)$; the construction in the case $u(b) = v(b)$ is similar.

Then $u = u(t)$, $v = v(t)$ are defined on $[a, b]$, $b - a = 2^{-n}$;

$$u(t) < v(t) \quad \text{on} \quad [a, b] ,$$

$$u(a) = v(a) , \ u'(a) = v'(a) ; \tag{35}$$

$$|u'(t)| , |u''(t)| , |v'(t)| , |v''(t)| \leq M_n ; \tag{36}$$

$$|u(t) - v(t)| , |u'(t) - v'(t)| \leq d_n \leq \epsilon_n . \tag{37}$$

Let $m = m_n > 0$ be an integer to be specified below. For $i = 0, 1, \dots, m$, put

$$u_i(t) = \frac{(m-i)u(t) + iv(t)}{m} = u(t) + [v(t) - u(t)]\frac{i}{m} \tag{38}$$

on $[a, b]$. Then $u_0(t) = u(t)$, $u_m(t) = v(t)$, and

$$u(t) \leq u_i(t) < u_{i+1}(t) \leq v(t) \quad \text{on} \quad (a, b] , \tag{39}$$

$$u_i(a) = u(a) , \quad u_i'(a) = u'(a) \tag{40}$$

for $i = 0, 1, \dots, m$. It is clear from (37) that

$$|u_h - u_i| \leq |h - i|\frac{d_n}{m} \leq d_n , \tag{41}$$

$$|u_i'| \leq M_n , \quad |u_h' - u_i'| \leq |h - i|\frac{d_n}{m} \leq d_n , \tag{42}$$

$$|u_i''| \leq M_n . \tag{43}$$

For $i = 0, 1, \ldots, m-1$ and $c = \dfrac{a+b}{2}$, put

$$v_i(t) = u_i(t)\sin^2\left(2^{n+1}\pi(t-c)\right) + u_{i+1}(t)\cos^2\left(2^{n+1}\pi(t-c)\right) \quad (44)$$

on$[c, b]$, so that

$$c - a = b - c = \frac{1}{2^{n+1}}$$

implies that

$$u_i(t) < v_i(t) < u_{i+1}(t) \quad (45)$$

on (c, b),

$$\left.\begin{array}{l} v_i = u_{i+1}\ ,\quad v_i' = u_{i+1}' \quad \text{at} \quad t = c \\ v_i = u_i\ , v_i' = u_i' \quad \text{at} \quad t = b \end{array}\right\}. \quad (46)$$

The relations(46) involving derivatives follow from

$$\begin{aligned} v_i' &= u_i'\sin^2\left(2^{n+1}\pi(t-c)\right) + u_{i+1}'\cos^2\left(2^{n+1}\pi(t-c)\right) \\ &\quad + 2^{n+1}\pi(u_i - u_{i+1})\sin\left(2^{n+2}\pi(t-c)\right)\ . \end{aligned} \quad (47)$$

From (47) and (41)–(43),

$$\left.\begin{array}{l} |v_i'| \le M_n + 2^{n+1}\pi\dfrac{d_n}{m} \\ |v_i''| \le M_n + \left(2^{2n+3}\pi + 2^{n+2}\right)\pi\dfrac{d_n}{m} \end{array}\right\}. \quad (48)$$

Also, by (44),

$$v_i - u_i = (u_{i+1} - u_i)\cos^2\left(2^{n+1}\pi(t-c)\right)\ ,$$

$$v_i - u_{i+1} = (u_i - u_{i+1})\sin^2\left(2^{n+1}\pi(t-c)\right)\ ,$$

so that (41)–(42) give

$$|v_i - u_h| \le \frac{d_n}{m}\ , \quad |v_i' - u_h'| \le \left(1 + 2^{n+1}\pi\right)\frac{d_n}{m} \quad (49)$$

for $h = i\ , i+1$.

Finally, let $m = m_n$ be chosen so large that

$$\left(2^{n+2} + 2^{2n+3}\pi\right)\pi\frac{d_n}{m} < \frac{\epsilon_{n+1}}{3} \,. \tag{50}$$

In order to obtain S_{n+1} from S_n, let the arcs $u = u_i(t)(i = 0, 1, \ldots, m)$ on $[a, c]$ and the arcs $v = v_h(t)(h = 0, 1, \ldots, m-1)$ on $[c, b]$ be inserted between $u = u(t)$, $v = v(t)$.

It is clear from (42)–(43), (48)–(50) that the analogues of (32) and (33) hold if n is replaced by $n+1$. Also the analogue of (34) follows from (42), (49)–(50).

This completes the construction of the sequence $S_0 \subset S_1 \subset \ldots$. It is clear that $S = \cup S_n$ is dense in the (t, u)-plane.

The *continuity* of $U = U(t, u)$, given by (30), will now be considered.

Let $p \geq n \geq 0$. The set of arcs S_n divide the plane into closed sets G of the form

$$G = \{(t, u) : \gamma \leq t \leq \delta \,,\ u^n(t) \leq u \leq v^n(t)\} \,,$$

where no point of S_n is interior to G;

$$u = u^n(t) \quad \text{and} \quad u = v^n(t) \quad \text{on} \quad [\gamma, \delta] \,,$$

$$\delta - \gamma = \frac{2}{2^n} \,,$$

are arcs each consisted of two arcs of S_n;

$$u^n = v^n \quad \text{at} \quad t = \gamma\,, \delta \ ;\ u^n < v^n \quad \text{on} \quad (\gamma, \delta) \,.$$

Let $(t_0, u_p) \in G \cap S_p$ and let (t^1, u^1) be any point of boundary of G.

The difference

$$\Delta = \left|U_p(t_0, u_p) - U_p(t^1, u^1)\right|$$

will be estimated.

Consider first the case that $p = n$. Then (t_0, u_p) is on the boundary of G, say $u_p = u^n(t_0)$.

Since

$$U_n(t, u^n(t)) = u^{n\,\prime}(t) \;,$$

it seen by (32) that

$$|U_n(t_0, u_n) - U_n(\gamma, u(\gamma))| \leq M_n |t_0 - \gamma| \leq \frac{2M}{2^n} \;.$$

Thus in the case that

$$p = n \;, \; \Delta_n \leq \frac{4M}{2^n} \;.$$

Let $p > n$. It can be supposed that $(t_0, u_p) \in S_p - S_{p-1}$.

Let

$$u_n = u^n(t_0) \quad \text{and} \quad u_n \leq u_{n+1} \leq \ldots \leq u_p \;,$$

where (t_0, u_j) is the highest point of the segment

$$t = t_0 \;, \; u_{j-1} \leq u \leq u_p$$

which is in S_j , $j = n+1, \ldots, p$.

Then, by

$$|U_p\,(t_0, u_{j+1}) - U_p(t_0, u_j)| \leq d_j + \epsilon_{j+1} \leq 2\epsilon_j \;.$$

Hence

$$|U_p(t_0, u_p) - U_p(t_0, u_n)| \leq 2 \sum_{j=n}^{\infty} \epsilon_j \;.$$

If this is combined with

$$\Delta_n \leq \frac{4M}{2^n} \;,$$

it follows that

$$\Delta_p \leq \eta_n \ , \tag{51}$$

where

$$\eta_n = \frac{4M}{2^n} + 2\sum_{j=n}^{\infty} \epsilon_j \ .$$

Consider now two points

$$(t_i, u_i) \ , \ i = 0, 1 \ , \text{ in } S_p \ , \ p \geq n \ .$$

Each of these points (t_i, u_i) is contained in the region $G = G_i$ of the type just considered. There exist points (t^i, u^i) on the boundary of G_i such that

$$|t^0 - t^1|^2 + |u^0 - u^1|^2 \leq |t_0 - t_1|^2 + |u_0 - u_1|^2 \ , \tag{52}$$

where, e.g. :

$$(t^0, u^0) = (t^1, u^1) \quad \text{if} \quad G_0 = G_1 \ .$$

Thus the above estimate for Δ_p implies

$$|U(t_0, u_0) - U(t_1, u_1)| \leq 2\eta_n + |U_n(t^0, u^0) - U_n(t^1, u^1)| \ . \tag{53}$$

Since $U_n = U_n(t, u)$ is uniformly continuous on S_n it follows from (51)– (53) that $U = U(t, u)$ *is uniformly continuous on* S.

Hence $U = U(t, u)$ has *a continuous extension*, denoted also by $U(t, u)$, on the (t, u)-plane. It will now be verified that the initial value problem

$$u' = U(t, u) \ , \ u(t_0) = u_0 \tag{54}$$

has the asserted property.

It is clear that any continuous arc $u = u(t)$ on an internal $[c, d]$ made up of subarcs of S is a solution of (54).

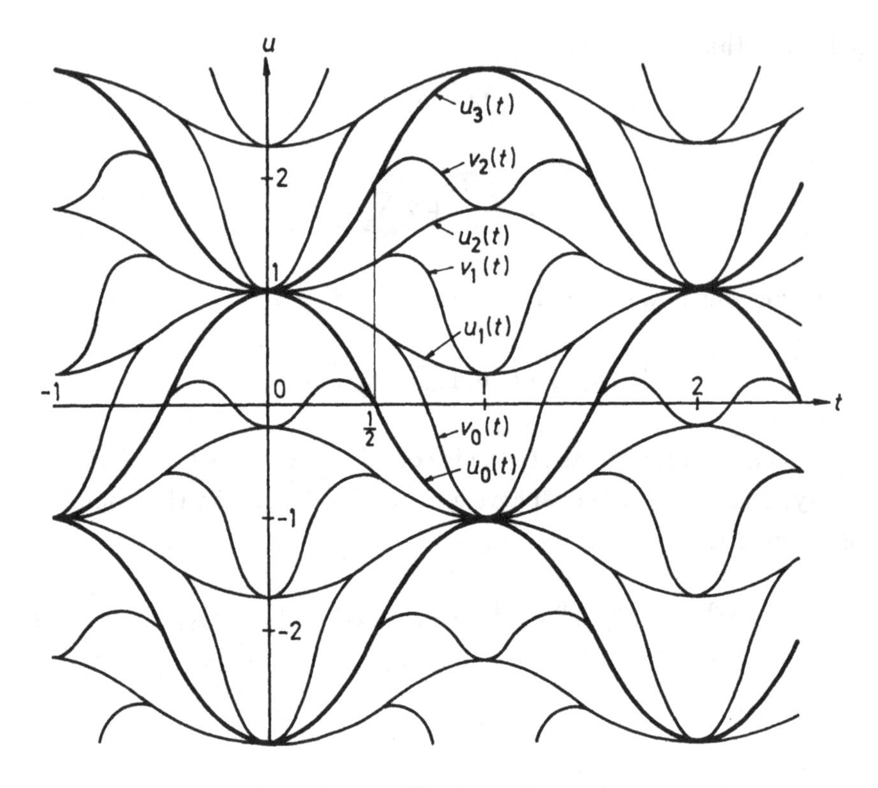

Fig. 5.5

Theorem (above) shows that if (t_0, u_0) is any point of the set G of the type just considered, then Problem (54) has a solution $u = u(t)$ over $[\gamma, \delta]$ satisfying

$$u = u^n = v^n \quad \text{at} \quad t = \gamma, \delta \ .$$

Such a solution can be continued to the left of $t = \gamma$ (right of $t = \delta$) ina *nonunique* manner by using arcs of S_n.

If n is sufficiently large, the interval $[\gamma, \delta]$ containing t_0 can be made arbitrarily small.

This completes the verification.

Remark: The above illustrates *how bad the situation* can become when we are looking for uniqueness of an initial value problem.

Note: The heavy curved lines of Figure 5.5 represent arcs of S_0. Heavy and light curved lines represent arcs of S_1 if $m_0 = 3$. The construction of the arcs of S_1, not in S_0, is indicated above: The arcs

$$u(t) = u_0(t)\ ,\ u_1(t)\ ,\ u_2(t)\ ,\ u_3(t) = v(t)$$

are defined on $[a,b] = [0,1]$, the arcs

$$v_0(t)\ ,\ v_1(t)\ ,\ v_2(t)$$

are defined on $[c,b] = [\frac{1}{2}, 1]$.

The sketch makes it clear how S_0 or S_1 divides the plane into sets G.

5.7 *A curve of the linear PDE of order two in two variables: $au_{xx} + 2bu_{xy} + cu_{yy} + du_x + eu_y + fu = g$, that is exceptional in the sense that the values of u, u_x, and u_y along the curve, together with the PDE, do not uniquely determine the values of u_{xx} , u_{xy}, and u_{yy} along the curve.*

Consider Γ a smooth curve in the xy-plane, given parametrically by the equations (**16**, p. 13–14, and **23**, p. 101–102):

$$x = x(s)\ ,\ y = y(s)\ ,\ s_1 < s < s_2\ . \tag{55}$$

Assume u, u_x, and u_y are specified along Γ as

$$u = F(s)\ ,\ u_x = G(s)\ ,\ u_y = H(s)\ . \tag{56}$$

Then

$$\left.\begin{aligned} \frac{du_x}{ds} &= u_{xx}\frac{dx}{ds} + u_{xy}\frac{dy}{ds} = \frac{dG}{ds} \\ \frac{du_y}{ds} &= u_{xy}\frac{dx}{ds} + u_{yy}\frac{dy}{ds} = \frac{dH}{ds} \end{aligned}\right\} \tag{57}$$

Equations (57) and PDE

$$au_{xx} + 2bu_{xy} + cu_{yy} + du_x + eu_y + fu = g \tag{58}$$

comprise three linear equations, which may be solved uniquely for the three unknowns u_{xx} , u_{xy}, and u_{yy} along Γ, unless the coefficient matrix

$$A = \begin{bmatrix} a & 2b & c \\ \frac{dx}{ds} & \frac{dy}{ds} & 0 \\ 0 & \frac{dx}{ds} & \frac{dy}{ds} \end{bmatrix}$$

has determinant

$$|A| = \begin{vmatrix} a & 2b & c \\ \frac{dx}{ds} & \frac{dy}{ds} & 0 \\ 0 & \frac{dx}{ds} & \frac{dy}{ds} \end{vmatrix}$$

of zero value; that is, unless

$$a\left(\frac{dy}{ds}\right)^2 - 2b\frac{dx}{ds}\frac{dy}{ds} + c\left(\frac{dx}{ds}\right)^2 = 0 . \tag{59}$$

Equation (59) is equivalant to the equation

$$a(dy)^2 - 2b\,dxdy + c(dx)^2 = 0 . \tag{60}$$

Equation (60) defined the characteristics of PDE (58) and is called *characteristics equation of PDE* (58).

Therefore *a characteristic curve* Γ of PDE (58) is an exceptional curve in the sense that the values of (56), together with PDE (58), do *not uniquely* determine the values of u_{xx} , u_{xy}, and u_{yy} along Γ. **Note:** The solution of the Cauchy problem for equation: $u_{xy} = f(x, y, u, u_x, u_y)$ is not unique.

5.8 *The Initial Value Problem:*

$$uu_x + u_t = 0 \ ,$$

$$u(x,0) = \begin{cases} 1 & , \quad 0 < x < 1 \\ 0 & , \quad x < 0 \ or \ x > 1 \end{cases}$$

admits two weak solutions.

Definition: Let Ω be the region: $x_1 < x < x_2$, $t_1 < t < t_2$ and suppose that in Ω

$$u = (u_1(x,t) \ , \ u_2(x,t) \ , \ \ldots \ , \ u_n(x,t)) = u(x,t)$$

solves the divergence-form first-order equation

$$\nabla \cdot \vec{V} = \frac{\partial}{\partial x}F(u) + \frac{\partial}{\partial t}G(u) = 0 \ , \tag{61}$$

where

$$\nabla = \left(\frac{\partial}{\partial x}, \frac{\partial}{\partial t}\right) \ , \ \vec{V} = \vec{V}(u) = (F(u), G(u)) \ .$$

Smooth ($\mathbf{C}^\infty$) functions $\varphi = \varphi(x,t)$ which vanish in a neighborhood and not simply on the boundary $\partial\Omega$ of Ω are called *test functions* on Ω. We say that $u = u(x,t)$ is *a weak solution of equation* (61) in Ω if

$$\iint\limits_{\Omega} \left[\frac{\partial \varphi}{\partial x}F(u) + \frac{\partial \varphi}{\partial t}G(u)\right] dxdt = 0 \tag{62}$$

holds for *all* test functions $\varphi = \varphi(x,t)$ on Ω.

Consider the Initial Value Problem:

$$uu_x + u_t = 0 \tag{63}$$

or equivalently (**16**, p. 71):

$$\frac{\partial}{\partial x}(u^2) + \frac{\partial}{\partial t}(u) = 0 \tag{63$'$}$$

(where $F = F(u) = u^2$, $G = G(u) = u$) with

$$u(x,0) = \begin{cases} 1 & , \quad 0 < x < 1 \\ \\ 0 & , \quad x < 0 \text{ or } x > 1 \, . \end{cases} \tag{64}$$

Problem (63) (or(63)′)–(64) admits *two weak solutions*. In fact, these solutions are the following

$$v = v(x,t) = \begin{cases} 0 & x < \frac{t}{2} \\ \\ 1 & \frac{t}{2} < x < 1 + \frac{t}{2} \\ 0 & x > 1 + \frac{t}{2} \end{cases}$$

and

$$w = w(x,t) = \begin{cases} 0 & x < 0 \\ \\ \frac{x}{t} & 0 < x < t \\ \\ 1 & t < x < 1 + \frac{t}{2} \\ \\ 0 & x > 1 + \frac{t}{2} \, . \end{cases}$$

5.9 *The Exterior Dirichlet Problem: $\Delta u = 0$ in $\Omega = \{x : r = |x| > 1\}$, $x = (x_1, x_2, \ldots, x_n) \in \mathbb{R}^n$, $|x| = (x_1^2 + x_2^2 + \ldots + x_n^2)^{\frac{1}{2}}$ with $u = 1$ on $S = \partial\Omega$: boundary of Ω, where neither $|u| \leq A$ in Ω, nor $u(x) \to 0$ uniformly as $|x| \to \infty$, has infinitely many solutions.*

Consider Problem (**14**, p. 240–244; **42**, p. 266, 288, 289):

$$\left.\begin{aligned} \Delta u = \sum_{i=1}^{n} \frac{\partial^2}{\partial x_i^2} u = 0 \quad &\text{in} \quad \Omega \\ u = 1 \quad \text{on} \quad &S \end{aligned}\right\} . \tag{65}$$

Problem (65) is satisfied by the one-parameter family of functions

$$v_{(\lambda)} = v_{(\lambda)}(r) = \begin{cases} \lambda + \dfrac{1-\lambda}{r^{n-2}}, & n > 2 \\ 1 + (1-\lambda)\ln r, & n = 2, \end{cases} \tag{66}$$

where $-\infty < \lambda < \infty$, and such that neither $|v_{(\lambda)}| \leq A$ in Ω, nor $v_{(\lambda)} \to 0$ uniformly, as $|x| \to \infty$.

Remarks: *The Dirichlet Problem for the half-plane $y > 0$ (and half-hyper-space $x_n > 0$) is not uniquely soluble* (i.e.: The Dirichlet Problem for Laplace equation: $\Delta u = 0$ in an unbounded domain cannot be solved in a unique way).

(i) For instance, the function

$$u(x, y) = y$$

is harmonic (i.e.: $\Delta u = u_{xx} + u_{yy} = 0$) in the whole plane of the variables x and y, in particular, in the half-plane $y > 0$, and it vanishes for $y = 0$ *without* vanishing identically for $y \geq 0$. Nevertheless, a class of functions can be indicated in which the considered problem has *at most one solution.* This class contains functions bounded in the half-plane $y > 0$ or in the half-hyper-space $x_n > 0$.

(ii) In the n-dimensional space, for $n \geq 3$, the *uniqueness* of the Exterior Problem does *not* follow even in the class of functions bounded in the exterior of a given closed domain. For instance,

$$u = u(x) = r^{2-n} - 1 \ , \ n > 2 \tag{67}$$

is harmonic and bounded in the exterior of the hyper-sphere D of radius 1 and center at the origin; moreover, $u(x) = 0$ on ∂D while u does *not* vanish identically in the exterior of D.

(iii) Consider $\Delta u = 0$ of (65) in Ω, the Γ- function $\Gamma(\frac{n}{2})$ and the surface area of the unit hyper-sphere S in n dimensions

$$\omega_n = \frac{2(\sqrt{\pi})^n}{\Gamma(\frac{n}{2})} \ . \tag{87}$$

Theorem:

If $u = u(x)$, $x = (x_1, x_2, \ldots, x_n)$, is a regular harmonic function in a domain G of x-space with boundary ∂G, then the function

$$v = v(x, y) = \frac{1}{r^{n-2}} \left(\frac{x_1}{r^2}, \frac{x_2}{r^2}, \ldots, \frac{x_n}{r^2} \right) \tag{88}$$

also satisfies the Laplace (or Potential) equation: $\Delta u = 0$ and is regular in the region G' obtained from G by inversion with respect to the unit hyper-sphere.

Note: Let the function $u = u(x)$ be regular and harmonic in a bounded domain G. If we invert G with respect to a unit hyper-sphere whose center, say the origin, lies in G, the interior of the G is carried into the exterior G' of the inverted boundary $\partial G'$.

The harmonic function (88) is then called *regular in the exterior region* G'.

That is, we define regularity *in a domain G extending the infinity*, as follows: We invert G with respect to the hyper-sphere with center outside of G, and thus transform G into a bounded domain G'.

The harmonic function $u = u(x)$ is called *regular at infinity* if G contains a neighborhood of the point at infinity and a value is assigned to the function u at the point at infinity such that $v = v(x)$ is regular in G'.

For instance,

$$u = \text{const} \tag{89}$$

is regular at infinity in the plane (that is, $n = 2$), but *not* in spaces ($n > 2$).

In space ($n > 2$), for arbitrary λ, the functions

$$u = \lambda + \frac{1-\lambda}{r} \ (= v_{(\lambda)} \quad \text{for} \quad n = 3) \tag{90}$$

are harmonic outside the unit sphere and have the boundary values

$$u = 1 \qquad \text{on the sphere.} \tag{91}$$

But

$$u = \frac{1}{r} \ (= v_{(0)} \quad \text{for} \quad n = 3) \tag{92}$$

is *the only function of this family* which is regular in the region exterior to the unit sphere. Similarly for any $n > 2$.

For any number n of dimensions *the only solutions of the potential (or Laplace) equation:* $\Delta u = 0$, that depends merely on the distance $r = |x - \xi|$ of the point x from a fixed point ξ (e.g., the origin), are (except for arbitrary multiplicative and additive constants) the functions

$$\gamma = \gamma(r) = \begin{cases} \dfrac{1}{(n-2)\omega_n} r^{2-n} & , \quad n > 2 \\[2ex] \dfrac{1}{2\pi} \ln \dfrac{1}{r} & , \quad n = 2 \end{cases} \tag{93}$$

which exhibit the so-called *characteristic singularity* for $r = 0$.

Every solution of $\Delta u = 0$ in G, $r_0 = |x - \xi| = \sum_{i=1}^{n}(x_i - \xi_i)^2$, of the form

$$\psi = \psi(x, \xi) = \gamma(r_0) + w \tag{94}$$

for ξ inside G and w regular, is called a *fundamental solution* with a singularity at the parameter point ξ.

5.10 *A System of Partial Differential Equations of first order for which elimination and partial differentiation do not lead to an equation of second order, or one equation of third order.*

Consider the system (**14**, p. 14)

$$\left.\begin{aligned} \frac{\partial u}{\partial x} + \frac{\partial v}{\partial y} &= -yu \\ \frac{\partial u}{\partial y} + \frac{\partial v}{\partial x} &= \quad yv \end{aligned}\right\} . \tag{95}$$

One obtains here *two equations of third order* for the (overdetermined) function u:

$$\frac{\partial}{\partial y}(y^2u + u_{yy} - u_{xx}) + u_x + yu = 0$$

and

$$\frac{\partial}{\partial x}(y^2u + u_{yy} - u_{xx}) + u_y - y\left(y^2u + u_{yy} - u_{xx}\right) = 0 .$$

5.11 *Solutions of Conservation Laws are not determined uniquely by their initial data.*

A Conservation Law for a single function $u = u(x,t)$ is an equation of the form (**14**, p. 148–151):

$$\frac{d}{dt}\int_{x_1}^{x_2} udx = f(u(x_2), x_2, t) - f(u(x_1)x_1, t) , \tag{96}$$

where f is a given function of u, x, and t.

Note: Equation (96) expresses the fact that the total quantity represented by the function u and contained in the interval (x_1, x_2) changes at the rate equal to *the "flux" f of u through the end points of (x_1, x_2)*. This is the form of those laws of physics which ignore dissipative mechanisms and thus express a "*phenomenon of conservation*". If u is a differentiable solution of (96), then the conservation law (96) is expressed by *the quasilinear differential equation*:

$$\frac{\partial u}{\partial t} = \frac{\partial f}{\partial u}\frac{\partial u}{\partial x} + \frac{\partial f}{\partial x} = \frac{\partial}{\partial x} f(u, x, t) , \tag{97}$$

obtained from (96) by differentiating with respect to x_1 and then setting $x_1 = x_2 = x$. Within the class of discontinuous solutions the conservation law (96) has a solution in the large, whereas the differential equation (97) does not.

For instance,

$$\frac{d}{dt}\int_{x_1}^{x_2} u dx = \frac{1}{2}u^2(x_2) - \frac{1}{2}u^2(x_1) . \tag{96)'$$

When differentiation is carried out, we obtain

$$\frac{\partial u}{\partial t} = u\frac{\partial u}{\partial x} . \tag{97)'$$

Dividing by u we have

$$\frac{u_t}{u} = (\ln u)_t = u_x . \tag{97)''$$

Denoting $\ln u$ by v; that is,

$$v = \ln u , \tag{98}$$

we can write (97)″ as the conservation law

$$\frac{d}{dt}\int_{x_1}^{x_2} v dx = e^{v(x_2)} - e^{v(x_1)} . \tag{96)''$$

The *jump relation for* (96)$'$ is

$$\frac{u_1 + u_2}{2} = -U \ , \tag{99}$$

where u_1 and u_2 denotes value of u on either side of the line of discontinuity.

The jump relation for (96)$''$ is

$$\frac{e^{v_1} - e^{v_2}}{v_1 - v_2} = -U \ . \tag{99$'$}$$

From these jump conditions we conclude: If $u = u(x,t)$ is a discontinuous solution of (96)$'$ then $v = \ln u$ is *not* a solution of (96)$''$.

Take the function

$$u = u(x,t) = \begin{cases} 1 & , \quad 2x < -t \\ \\ 0 & , \quad -t < 2x \ . \end{cases} \tag{100}$$

This function (100) is a discontinuous solution of (96)$'$, since on the either side of the line

$$2x = -t \ , \tag{101}$$

u is a constant and thus a smooth solution of equation (96)$'$, and across the line of discontinuity (101) the jump relation (99) holds.

Consider

$$u' = u'(x,t) = \begin{cases} 1 & , \quad x < -t \\ \\ -\dfrac{x}{t} & , \quad -t < x < 0 \ . \\ \\ 0 & , \quad 0 < x \end{cases} \tag{102}$$

This new function (102) is continuous for t positive and satisfies the differential equation everywhere except on the lines

$$x = 0 \quad \text{and} \quad x = -t \ . \tag{103}$$

From this one can easily show by integration that u' is a continuous solution of (96)$'$.

The two solutions u and u' have the same value at $t = 0$.

Remark:

It is possible to show that for arbitrarily prescribed initial data there exist *uncountably many discontinuous solutions* with the same prescribed data.

Among all this discontinuous solutions with the same initial values there is *only* one physical significance. This solution is called *permissible*

MATHEMATICAL PRINCIPLE (*CRITERION*): A discontinuous solution is permissible if every line of discontinuity is crossed by the forward drawn characteristics issuing from either side.

Analytically this condition means that for a permissible discontinuity

$$-a(u_L) \geq U \geq -a(u_R) \tag{103}$$

where u_L and u_R denote the value of u to the left and right of the line of discontinuity, and U denotes the velocity of the propagation of the discontinuity.

5.12 *The Solution of Equation : $u_{xx} = f(u)$ is not unique on the interval : $0 \leq x \leq 2\pi$, where $f(u)$ is a bounded function : $f(u) = -u$, $-1 \leq u \leq 1$.*

Consider equation (**14**, p. 369):

$$u_{xx} = f(u) \tag{104}$$

where $0 \leq x \leq 2\pi$ and $f(u)$ bounded such that $f(u) = -u$ for $-1 \leq u \leq 1$.

The functions

$$u = \lambda \sin x \ , \ |\lambda| \leq 1 \ , \tag{105}$$

are solutions of (104) vanishing at the end points.

Note: In general, if the domain is *not* sufficiently small the solution of equation

$$\Delta u = f(x, u) \tag{106}$$

need *not* be unique.

5.13 *A homogeneous Dirichlet problem for an elliptic system of Partial Differential equations with infinitely many solutions.*

Consider the elliptic system of partial differential equations (**7**, p. 46, 63, 202–203)

$$\left.\begin{aligned} u_{xx} - u_{yy} - 2v_{xy} &= 0 \\ v_{xx} - v_{yy} + 2u_{xy} &= 0 \end{aligned}\right\} . \tag{107}$$

Then consider the homogeneous Dirichlet problem with boundary conditions

$$u(t) = 0 \ , \ v(t) = 0 \ , \ |t| = 1 \tag{108}$$

for this system in the circle $|z| < 1$, $z = x + iy$.

Note: Let D be the domain in a 2-dimensional space $\mathbb{R}^2$ with boundary $S(= \partial D)$. By *a function of class* $\mathbf{C}^0(\bar{D})$, $\bar{D} = D \cup S$, is meant a one-valued function continuous in $\bar{D}$.

For *Dirichlet Problem*: To find a function u of class $\mathbf{C}^0(\bar{D})$ which is harmonic (i.e.: $\Delta u = 0$) in D and satisfies the boundary condition $u = g$ on S, where $g = g(x, y)$ is an arbitrary continuous function defined on S, we can prove that all *regular solutions of system* (107) (i.e.: Real functions $u = u(x, y)$ and $v = v(x, y)$ defined in $|z| < 1$ and continuous together with their partial derivatives involved in (107) which turn equations (107) into identities) in a simply-connected domain can be obtained from the formula

$$u(x, y) + iv(x, y) = \bar{z}\varphi(z) + \psi(z) \ , \tag{109}$$

where $\varphi = \varphi(z)$ and $\psi = \psi(z)$ are arbitrary analytic functions of the complex variable z.

In fact, system (107) can be written in the following form

$$w_{\bar{z}\bar{z}} = 0 \,, \tag{110}$$

where

$$z = x + iy \,, \; \bar{z} = x - iy \,, \; w = u + iv \,.$$

Therefore representation (109) follows.

Employing now general expression (109) for the solutions of (107) we can show that the Problem (107)–(108) in $|z| < 1$ possesses *infinitely many solutions* $u = u(x,y)$, $v = v(x,y)$ satisfying

$$u(x,y) + iv(x,y) = (1 - z\bar{z})\psi(z) \,, \tag{111}$$

where $\psi = \psi(z)$ is an arbitrary analytic function in the circle $|z| < 1$.

In fact, formula (109) implies that

$$\varphi(t) + t\psi(t) = t\,[f_1(t) + if_2(t)] \tag{112}$$

on the circle.

It follows that in case

$$f_1(t) = f_2(t) = 0 \tag{113}$$

there must be

$$\varphi(t) = -t\psi(t) \,, \; |t| = 1 \,. \tag{114}$$

Therefore, by virtue of the uniqueness theorem for analytic functions, we must have

$$\varphi(z) = -z\psi(z) \tag{115}$$

everywhere within the circle $|z| \leq 1$.

Hence, the Problem (107)–(108) possesses *infinitely many linearly independent solutions* $u = u(x,y)$, $v = v(x,y)$ *satisfying* (111).

5.14 *The Problem :* $u_{xx}-u_{tt}=0$, $u(x,t)=f_1(x)$, $\frac{\partial u}{\partial n}\left(=\frac{u_x+u_t}{\sqrt{2}}\right)$ $=\varphi_1(x)$ *on line* $x-t=0$, *with* $f_1'(x)=\sqrt{2}\varphi_1(x)$, *has infinitely many regular solutions* u.

Consider the equation of oscillation of a string (**7**, p. 66, 75, 77, 224):

$$u_{xx}-u_{tt}=0\ , \tag{116}$$

with the normal detiovative

$$\frac{\partial u}{\partial n}=\frac{u_x+u_t}{\sqrt{2}}\ , \tag{117}$$

the data

$$u=f_1(x)\ ,\ \frac{\partial u}{\partial n}=\varphi_1(x) \tag{118}$$

on the line

$$x-t=0 \tag{119}$$

and the *condition*

$$f_1'(x)=\sqrt{2}\,\varphi_1(x)\ . \tag{120}$$

The general solution of equation (116) is of the form

$$u=u(x,t)=f(x+t)+\varphi(x-t)\ , \tag{121}$$

where f and φ are arbitrary twice continuously differentiable functions.

From (117)–(120) we conclude that

$$\left.\begin{aligned} f(2x)+\varphi(0)&=f_1(x)\\ \sqrt{2}f'(2x)&=\varphi_1(x)\end{aligned}\right\}. \tag{122}$$

Therefore

$$\left.\begin{aligned} f(x)&=f_1(\frac{x}{2})-\varphi(0)\\ f'(x)&=\frac{1}{\sqrt{2}}\,\varphi_1(\frac{x}{2})\end{aligned}\right\}. \tag{123}$$

Consequently the Problem (116)–(119) is solvable only when condition (120) holds. Using condition (120), the solution of Problem (116)–(119) is given by the formula

$$u = u(x,t) = f_1\left(\frac{x+t}{2}\right) - \varphi(0) + \varphi(x-t) . \qquad (124)$$

Hence there are infinitely many solutions u.

5.15 *The Problem : $u_{xx} - u_{tt} = 0$, $u(x,0) = \varphi(x)$ for $0 \le x < \infty$, $u(0,t) = \psi(t)$ for $0 \le t < \infty$ in the first quadrant of the xt-plane, with $\varphi(0) = \psi(0)$ and $\varphi''(0) = \psi''(0)$ has infinitely many regular solutions u.*

Consider the equation (116), the data

$$\left.\begin{array}{ll} u(x,0) = \varphi(x) , & 0 \le x < \infty \\ u(0,t) = \psi(t) , & 0 \le t < \infty \end{array}\right\} \qquad (125)$$

in the first quadrant of the xt-plane, and the *conditions* (**7**, p. 79, 227):

$$\varphi(0) = \psi(0) , \ \varphi''(0) = \psi''(0) . \qquad (126)$$

In fact, the corresponding homogeneous problem possesses non-trivial solutions of the form

$$u = u(x,t) = \begin{cases} \omega\left(\frac{x+t}{2}\right) - \omega\left(\frac{x-t}{2}\right) , & x \ge t \\ \omega\left(\frac{x+t}{2}\right) - \omega\left(\frac{t-x}{2}\right) , & x \le t , \end{cases} \qquad (127)$$

where ω is an arbitrary twice continuously differentiable function satisfying conditions

$$\omega'(0) = \omega''(0) = 0 . \qquad (128)$$

Thus Problem (116), (125)–(126) has infinitely many solutions u.

5.16 *The Problem:* $u_{xx} - u_{tt} = 0$, $u(0,t) = \varphi(t)$ *and* $u(x, x/2) = \psi(x)$ *for* $t \geq 0$, $x \geq 0$ *in the angle-bounded domain* D *by the straight lines* $x = 0$, $t = x/2$, $t \geq 0$, $x \geq 0$, *with* $\varphi(0) = \psi(0)$ *and* $\varphi''(0) = \psi''(0)$ *has infinitely many solutions* u.

Consider the equation (116), the data

$$\left.\begin{aligned} u(0,t) &= \varphi(t) \ , \ t \geq 0 \\ u\left(x, \frac{x}{2}\right) &= \psi(x) \ , \ x \geq 0 \end{aligned}\right\} \tag{129}$$

in the angle-bounded domain D by the straight lines (**7**, p. 80, 227–228):

$$x = 0 \ , \ t = \frac{x}{2} \ , \ t \geq 0 \ , \ x \geq 0 \ , \tag{130}$$

and *conditions*

$$\varphi(0) = \psi(0) \ , \ \varphi''(0) = \psi''(0) \ . \tag{131}$$

In fact, the corresponding homogeneous problem possesses non-trivial solutions of the form

$$u = u(x,t) = \begin{cases} \omega\left(\dfrac{x+t}{2}\right) - \omega\left(\dfrac{3}{2}(x-t)\right) & , \quad \dfrac{x}{2} \leq t \leq x \\ \omega\left(\dfrac{x+t}{2}\right) - \omega\left(\dfrac{t-x}{2}\right) & , \quad t \geq x \ , \end{cases} \tag{132}$$

where ω is an arbitrary twice continuously differentiable function satisfying conditions

$$\omega'(0) = \omega''(0) = 0 \ . \tag{133}$$

Thus Problem (116), (129)–(131) has infinitely many solutions u.

5.17 *The Problem:* $u_{xx} - u_{tt} = 0$, $u(x,x) = \varphi(x)$ *for* $0 \leq x \leq \infty$, $(u_x + u_t)|_{t=x} = \psi(x)$ *for* $0 \leq x < \infty$, *with* $\varphi'(x) = \psi(x)$ *has infinitely many solutions* u.

Consider the equation (116), the data

$$\left.\begin{array}{l} u(x,x)=\varphi(x)\,, \quad 0\le x<\infty \\ \left(\dfrac{\partial}{\partial x}+\dfrac{\partial}{\partial t}\right)u(x,x)=\psi(x)\,, \quad 0\le x<\infty \end{array}\right\}, \tag{134}$$

and *condition* (**7**, p. 82, 231):

$$\varphi'(x)=\psi(x)\,,\ 0\le x<\infty\,. \tag{135}$$

In fact, the solution of Problem (116), (134)–(135) has the form

$$u=u(x,t)=\varphi\left(\frac{x+t}{2}\right)-f(x-t)\,, \tag{136}$$

where f is an arbitrary twice continuously differentiable function satisfying condition

$$f(0)=\varphi(0)\,. \tag{137}$$

Thus, Problem (116), (134)–(135) has infinitely many solutions u.

5.18 *The Problem:* $u_{xx}-u_{tt}=0\,,\ u(a,t)=u(b,t)=0$ *for* $t>0$ *in the half-strip* $a<x<b\,,\ t>0$ *has infinitely many solutions* u.

See: **7**, p. 109, 258.

Consider equation (116), and the data

$$u(a,t)=u(b,t)=0\,,\ t>0 \tag{138}$$

in the half-strip $a<x<b\,,\ t>0$.

In fact, problem (116), (138) possesses infinitely many solutions of the form

$$u_n=u_n(x,t)=\left[a_n\cos\left(\frac{\pi n}{b-a}t\right)+b_n\sin\left(\frac{\pi n}{b-a}t\right)\right]\cdot\sin\frac{\pi n}{b-a}(x-b)\,,$$

$n = 1, 2, 3, \ldots$, where a_n and b_n are arbitrary real constants.

5.19 *The Cauchy problem :* $y^2 u_{xx} + y u_{yy} + \frac{1}{2} u_y = 0$ *with data* $u(x,0) = \varphi(x)$, $u_y(x,0) = 0$ *for* $0 < x < 1$, $y < 0$ *has infinitely many solutions* u.

Consider equation (**7**, p. 90, 243):

$$y^2 u_{xx} + y u_{yy} + \frac{1}{2} u_y = 0 \tag{139}$$

with data

$$\left.\begin{aligned} u(x,0) &= \varphi(x) \\ u_y(x,0) &= 0 \end{aligned}\right\} \tag{140}$$

for $0 < x < 1$, $y < 0$.

In fact, the transformation of variables

$$\xi = x + \frac{2}{3}(-y)^{\frac{3}{2}} \ , \ \eta = x - \frac{2}{3}(-y)^{\frac{3}{2}} \tag{141}$$

(which is non-singular for $y < 0$) reduces the given equation to the form

$$u_{\xi\eta} = 0 \ . \tag{142}$$

Therefore it follows that the general solution of the original equation (139) is

$$u = u(x,y) = f_1\left(x + \frac{2}{3}(-y)^{3/2}\right) + f_2\left(x - \frac{2}{3}(-y)^{3/2}\right) \ , \tag{143}$$

where $f_1 = f_1(t)$ and $f_2 = f_2(t)$ are arbitrary twice continuously differentiable functions so that

$$f_1(x) + f_2(x) = \varphi(x) \tag{144}$$

and

$$\lim_{y \to 0^-} (-y)^{1/2} \cdot \left[-f_1'\left(x + \frac{2}{3}(-y)^{3/2}\right) + f_2'\left(x - \frac{2}{3}(-y)^{3/2}\right)\right] = 0 \tag{145}$$

hold for $0 < x < 1$, $y < 0$.

In this case (143)–(144) give the solution

$$u = u(x,y) = \varphi\left(x + \frac{2}{3}(-y)^{3/2}\right) - f_2\left(x + \frac{2}{3}(-y)^{3/2}\right) + f_2\left(x - \frac{2}{3}(-y)^{3/2}\right),$$

where $f_2 = f_2(t)$ is an arbitrary twice continuously differentiable function.

Therefore problem (139)–(140) has infinitely many solutions u.

Remark: If we assume data

$$\left.\begin{aligned} u(x,0) &= \varphi(x) \\ u_y(x,0) &= \psi(x) \ (\neq 0) \end{aligned}\right\}, \qquad (140)'$$

then

$$\lim_{y\to 0-} (-y)^{1/2}\left[-f_1'\left(x + \frac{2}{3}(-y)^{3/2}\right) + f_2'\left(x - \frac{2}{3}(-y)^{3/2}\right)\right] = \psi(x),$$

contradicting hypothesis $\psi(x) \neq 0$.

Therefore Problem (139)–(140)′ does *not* have any solution.

5.20 *The Problem* : $\dfrac{\partial^3 u}{\partial x^3} - \dfrac{\partial^3 u}{\partial x \partial y^2} = 0$ *with* $u(x,0) = \varphi_1(x)$, $u_y(x,0) = \varphi_2(x), u_{yy}(x,0) = \varphi_3(x)$ *and condition* $\varphi_1''(x) = \varphi_3(x) - f_3''(0)$, *has infinitely many solutions* u.

Consider equation (**7**, p. 92, 246):

$$\frac{\partial^3 u}{\partial x^3} - \frac{\partial^3 u}{\partial x \partial y^2} = 0 \qquad (146)$$

with data

$$\left.\begin{aligned} u(x,0) &= \varphi_1(x) \\ u_y(x,0) &= \varphi_2(x) \\ u_{yy}(x,0) &= \varphi_3(x) \end{aligned}\right\}, \qquad (147)$$

and *condition*

$$\varphi_1''(x) = \varphi_3(x) - f_3''(0) . \tag{148}$$

In fact, the general solution of the equation (146) has the form

$$u = u(x,y) = f_1(x+y) + f_2(x-y) + f_3(y) , \tag{149}$$

where f_1 , f_2 , f_3 are arbitrary sufficiently smooth functions. This is justified as follows:

Write equation (146) in the form

$$\frac{\partial}{\partial x}\left(\frac{\partial^2 u}{\partial x^2} - \frac{\partial^2 u}{\partial y^2}\right) = 0 . \tag{146'}$$

It follows that

$$u_{xx} - u_{yy} = -f_3''(y) , \tag{150}$$

where $f_3 = f_3(y)$ is an arbitrary twice continuously differentiable function. Since f_3 is a particular solution of equation (150) and since the expression $f_1(x+y) + f_2(x-y)$ is the general solution of the homogeneous equation corresponding to (150), where f_1 and f_2 are arbitrary functions possessing continuous partial derivatives up to the third order, we conclude that (149) is the general solution of the equation (146).

Employing the formula (149) we obtain the system of equations

$$\left.\begin{array}{l} f_1(x) + f_2(x) + f_3(0) = \varphi_1(x) \\ f_1'(x) + f_2'(x) + f_3'(0) = \varphi_2(x) \\ f_1''(x) + f_2''(x) + f_3''(0) = \varphi_3(x) \end{array}\right\} . \tag{151}$$

Note: System (151) shows that the Problem (146)–(147) *cannot have a solution* if

$$\varphi_1''(x) \neq \varphi_3(x) - f_3''(0) \tag{148'}$$

(that is, if (148) does *not* hold).

If condition (148) holds, then

$$f_1(x) = \frac{1}{2}\varphi_1(x) + \frac{1}{2}\int_0^x \varphi_2(t)dt - \frac{1}{2}f_3'(0)x - \frac{1}{2}f_3(0) + c \tag{152}$$

and

$$f_2(x) = \frac{1}{2}\varphi_1(x) - \frac{1}{2}\int_0^x \varphi_2(t)dt + \frac{1}{2}f_3'(0)x - \frac{1}{2}f_3(0) - c\,, \tag{153}$$

where c is an arbitrary constant.

Therefore the solution of Problem (146)–(147) is given by the formula

$$u = u(x,y) = \frac{1}{2}\varphi_1(x+y) + \frac{1}{2}\varphi_1(x-y) + \frac{1}{2}\int_{x-y}^{x+y} \varphi_1(t)dt + f_3(y) - f_3'(0)y - f_3(0)\,. \tag{149$'$}$$

Thus Problem (146)–(147) has infinitely many solutions of the form (149)′.

5.21 *A Nonlinear Example of an Elliptic equation for which the uniqueness fails.*

Consider the Problem (**58**, p. 58–59):

$$\left.\begin{aligned} \frac{\partial u}{\partial t} &= \Delta u - f(u) \quad \text{in} \quad D \times [0,T) \\ u &= 0 \quad \text{on} \quad \partial D \times [0,T) \\ u(x,0) &= u_0(x) \\ T>0\;,\; D &: \text{given domain},\; x \in D\;,\; t \in [0,T) \end{aligned}\right\}, \tag{154}$$

where the point function $f = f(u)$ is assumed to satisfy the inequality

$$uf(u) \geq k|u|^{2-\epsilon} , \tag{155}$$

for $0 < \epsilon < 1$ and some constant k.

Consider now the following $\mathbf{L}_p$-integral

$$\Phi_p(t) = \int_D u^{2p} dx , \tag{156}$$

p : a positive integer.

Now

$$\frac{d\Phi_p}{dt} = 2p \int_D u^{2p-1} \frac{\partial u}{\partial t} dx , \tag{157}$$

or

$$\frac{d\Phi_p}{dt} = -2p(2p-1) \int_D u^{2p-2} |\text{grad} u|^2 dx - 2p \int_D u^{2p-1} \mathcal{F}(u) dx , \tag{157$'$}$$

where $\mathcal{F} = \mathcal{F}(u)$ is a given function of u.

Employing (155) we have, if we neglect the first term on the right of (157)$'$,

$$\begin{aligned} \frac{d\Phi_p}{dt} &\leq -2p \int_D |u|^{2p-\epsilon} dx \\ &\leq -\frac{2p}{(u^*(\eta))^{\epsilon}} \Phi_p , \end{aligned} \tag{158}$$

where again

$$u^*(\eta) = \max_{x \in D} |u(x, \eta)| . \tag{159}$$

An integration of (158) yields

$$\Phi_p(t) \leq \exp\left(-2p \int_0^t \frac{d\eta}{(u^*(\eta))^{\epsilon}} \right) \Phi_p(0) . \tag{160}$$

We now raise both sides of (160) to the $\frac{1}{2p}$ power and let $p \to \infty$ to obtain

$$u^*(t) \leq \exp\left(-2\int_0^t \frac{d\eta}{(u^*(\eta))^\epsilon}\right) u^*(0) \,, \tag{161}$$

or (provided $u^*(t) > 0$),

$$(u^*(0)^{-\epsilon} \leq (u^*(t))^{-\epsilon} \exp\left(-2\epsilon \int_0^t \frac{d\eta}{(u^*(\eta))^\epsilon}\right) . \tag{161$'$}$$

Integrating both sides of (161)$'$ from 0 to t, we obtain

$$(u^*(0)^{-\epsilon} t \leq \frac{1}{2\epsilon}\left[1 - \exp\left(-2\epsilon \int_0^t \frac{d\eta}{(u^*(\eta))^\epsilon}\right)\right] \leq \frac{1}{2\epsilon} \,, \tag{162}$$

which clearly cannot hold for all t.

Thus if a solution $u(t)$ exists, then it must tend to zero in a finite time t_1 which satisfies

$$t_1 \leq \frac{1}{2\epsilon}(u^*(0))^\epsilon \,. \tag{163}$$

This clearly gives an example of the nonuniqueness backward in time for the nonlinear equation of (154).

5.22 *Non-uniqueness of the Cauchy heat problem.*

Consider heat equation (**18**, p. 349):

$$\frac{\partial u}{\partial t} = \frac{\partial^2 u}{\partial x^2} \,, -\infty < x < \infty \,, \tag{164}$$

with data

$$u(0) = 0 \,. \tag{165}$$

The function

$$u = u(x,t) = \sum_{n=0}^{\infty} \frac{g^{(n)}(t)}{(2n)!} x^{2n} \tag{166}$$

where

$$g = g(t) = \begin{cases} e^{-t^{-2}} & , \quad t > 0 \\ 0 & , \quad t \leq 0 \, , \end{cases} \tag{167}$$

is a formal solution of equation (164). The function g is infinitely differentiable in $-\infty < t < \infty$. The series of (166) is uniformly convergent (together with all its derivatives) on compact subsets of the (x,t)-plane. Moreover, there exists a constant θ, $0 < \theta < 1$, such that

$$|u(x,t)| \leq \exp\left(\frac{x^2}{\theta t} - \frac{1}{2t^2}\right), \qquad -\infty < x < \infty \, , t > 0 \, . \tag{168}$$

Note that

$$\frac{x^2}{\theta t} - \frac{1}{2t^2} \leq \frac{x^4}{2\theta^2}, \qquad -\infty < x < \infty \, , t > 0 \, . \tag{169}$$

Consider

$$\begin{aligned} \mathbf{E} = \mathbf{C}_{a,4} & \\ = & \text{ space of all functions } u \text{ defined in } -\infty < x < \infty \\ & \text{ and such that norm} \end{aligned} \tag{170}$$

$$\|u\| = \sup_{-\infty < x < \infty} |u(x)|(1 + |x|)\exp(-ax^4) \quad (< \infty)$$

with $a > \dfrac{1}{2\theta^2}$.

In addition, consider the operator A defined by the formula

$$Au = u'' \tag{171}$$

with

$D(A) = \{$all twice differentiable u in E with $u'' \in E\}$.

It is *not* difficult to see that the function

$$t \to u(\cdot, t) \tag{172}$$

is finitely differentiable in the sense of the norm of E and solves

$$u'(t) = Au(t) \tag{173}$$

for all t.

Since (165) holds, solutions of equation (164) are *not* uniquely determined by initial data (165).

6. STABILITY

6.1 *An unstable nonlinear plane autonomous first-order system which is asymptotically attractive.*

The answer is given by Thomas Brown in the following way (**5**, p. 121–122, example 6):

Preliminaries: Let $a = (a_1, a_2)$ be a critical point of the autonomous system

$$\frac{dx}{dt} = X(x,y) \ , \ \frac{dy}{dt} = Y(x,y) \ , \tag{$*$}$$

so that

$$X(a_1, a_2) = 0 \ , \ Y(a_1, a_2) = 0 \ . \tag{1}$$

The critical point a is called :

(i) *Stable* when, given any $\epsilon > 0$, there exists a $\delta = \delta(\epsilon) > 0$ so small that, if

$$|x(0) - a_1| < \delta \ , \ |y(0) - a_2| < \delta \tag{2}$$

then

$$|x(t) - a_1| < \epsilon \ , \ |y(t) - a_2| < \epsilon \tag{3}$$

for all $t > 0$ and all orbits $x = x(t)\,,\ y = y(t)$,

(ii) *Asymptotically Attractive* when, for some $\delta > 0$ relations (2) imply

$$\lim_{t\to\infty} |x(t) - a_1| = 0 \ , \ \lim_{t\to\infty} |y(t) - a_2| = 0 \tag{4}$$

for all orbits: $x = x(t)\,,\ y = y(t)$.

(iii) *Strictly Stable* when, it is stable and asymptotically attractive,

(iv) *Neutrally Stable* when it is *not* asymptotically attractive,

(v) *Unstable* when, it is *not* stable.

Example (due to Thomas Brown):

Consider plane regions:

D_1: the lower half-plane $y \leq 0$;
D_2: the locus $x^2 + y^2 \leq 2|x|$, consisting of the discs $(x \pm 1)^2 + y^2 \leq 1$;
D_3: the half-strip $|x| \leq 2$, $y > 0$, exterior to D_2;
D_4: the locus $|x| > 2$, $y > 2$; as depicted in the following figure.

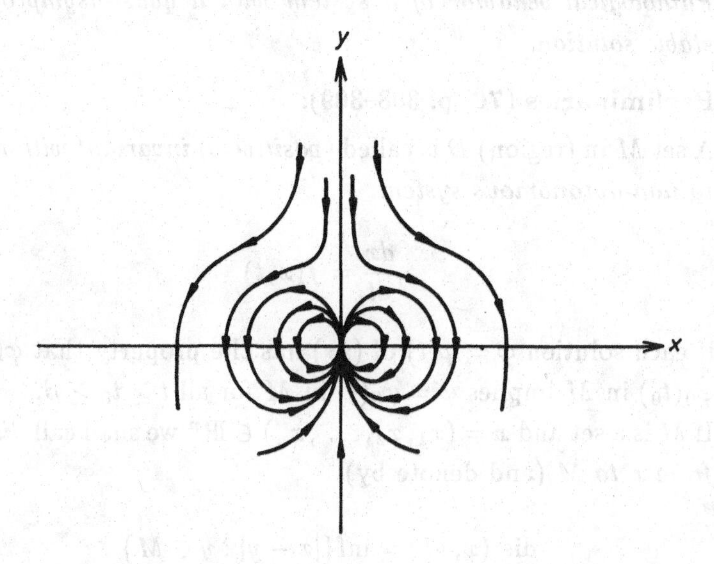

Fig. 6.1

The system

$$\frac{dx}{dt} = \begin{cases} 2xy & \text{on} \quad D_1 \cup D_2 \cup D_3 \\ \dfrac{2xy}{3 - 4|x|} & \text{on} \quad D_4 \, , \end{cases} \tag{5}$$

$$\frac{dy}{dt} = \begin{cases} y^2 - x^2 & \text{on} \quad D_1 \cup D_2 \, , \\ 4|x| - y^2 - 3x^2 & \text{on} \quad D_3 \, , \\ \dfrac{4|x| - y^2 - 3x^2}{3 - \dfrac{4}{|x|}} & \text{on} \quad D_4 \end{cases} \tag{6}$$

is *Unstable* , yet

$$\lim_{t\to\infty} |x(t)| = 0\ ,\ \lim_{t\to\infty} |y(t)| = 0 \tag{3)*}$$

for all orbits : $x = x(t)$, $y = y(t)$.

6.2 *Pathological behavior of a system with a quasi-asymptotically stable solution.*

Preliminaries (**76**, p. 308–309):

(**1**). A set M in (region) D is called (*positively*) *invariant with respect to non-autonomous system*

$$\frac{dx}{dt} = f(x,t) \tag{**}$$

if each solution $\varphi = \varphi(t)$ of (**) has the property that $\varphi(t_0) = \varphi_0(t_0)$ in M implies $\varphi = \varphi_0(t)$ in M for all $t \geq t_0 \geq \beta$.

(**2**). If M is a set and $x = (x_1, x_2, \ldots, x_n) \in \mathbb{R}^n$ we shall call *distance from x to M* (and denote by)

$$\operatorname{dist}(x, M) = \inf\{|x - y| : y \in M\}\ . \tag{7}$$

Let M be invariant with respect to (**). Then M is called

(i) *Stable* if, given any $\epsilon > 0$ and $t_0 \geq \beta$, there exists a $\delta > 0$ (which may depend on both t_0 and ϵ) so that whenever a solution $\varphi = \varphi(t)$ of (**) satisfies

$$\operatorname{dist}(\varphi(t_0), M) < \delta\ , \tag{2)'}$$

then $\varphi(t)$ exists and

$$\operatorname{dist}(\varphi(t), M) < \epsilon \tag{3)'}$$

for all $t_0 \leq t < \infty$,

(ii) *Asymptotically Stable* if, it is stable and furthermore for given $t_1 \geq \beta$ there is a $\delta_1 > 0$ such that whenever a solution $\varphi = \varphi(t)$ of (**) satisfies $|\varphi(t_1)| < \delta_1$ (or whenever $\text{dist}(\varphi(t_1), M) < \delta_1$) then

$$\lim_{t \to \infty} \text{dist}(\varphi(t), M) = 0 . \qquad (4)'$$

Note: In the case of an autonomous system

$$\frac{dx}{dt} = f(x) \qquad [**]$$

one sometimes attaches the phrase "*in the sense of Poincaré*" to definitions (i)–(ii) above.

Let $\psi = \psi(t)$ be a solution of (**) which exists for $\beta \leq t < \infty$. Then ψ is called

(i)′. *Stable* if, given any $\epsilon > 0$ and $t_0 \geq \beta$, there exists a $\delta > 0$ such that whenever a solution $\varphi = \varphi(t)$ of (**) satisfies

$$|\varphi(t_0) - \psi(t_0)| < \delta , \qquad (2)''$$

then $\varphi(t)$ exists and

$$|\varphi(t) - \psi(t)| < \epsilon \qquad (3)''$$

for all $t_0 \leq t < \infty$,

(ii)′. *Asymptotically Stable* if, it a stable and if, for given any $t_1 \geq \beta$ there is a $\delta_1 > 0$ such that whenever a solution $\varphi = \varphi(t)$ of (**) satisfies

$$|\varphi(t_1) - \psi(t_1)| < \delta_1 ,$$

then

$$\lim_{t \to \infty} |\varphi(t) - \psi(t)| = 0 . \qquad (4)''$$

Notes:

I. Stability as defined in (i)′–(ii)′ is sometimes designated as being "*in the sense of Lyapunov*".

II. If a solution (or invariant set) is stable and if every solution approaches it as $t \to \infty$, then it is called ***globally asymptotically stable.***

III. If a solution (or invariant set) is *not* stable, then it is called *unstable.*

IV. A solution $\psi = \psi(t)$ of $(**)$ which is unstable, even though every other solution $\varphi = \varphi(t)$ approaches it in the sense that $(4)''$ holds :

$$\lim_{t \to \infty} |\varphi(t) - \psi(t)| = 0 \ ,$$

is called *Quasi-asymptotically stable.*

Example : Consider the plane autonomous system

$$\left.\begin{aligned} \frac{dx}{dt} &= x - y + \frac{xy - x^3 - xy^2}{\sqrt{x^2 + y^2}} \\ \frac{dy}{dt} &= x + y - \frac{x^2 + x^2 y + y^3}{\sqrt{x^2 + y^2}} \end{aligned}\right\} . \tag{8}$$

In fact, the critical point (1,0) is a quasi-asymptotically stable solution of the system (8). It is clear this if we use polar coordinates.

Remarks:

(i). A quasi-asymptotically stable cruising attitude is an undesirable feature in an aircraft, because disturbances due to turbulence could be amplified.

(ii). A phase portrait for a system with a quasi-asymptotically stable solution $x = 0$, $y = 0$ is the following:

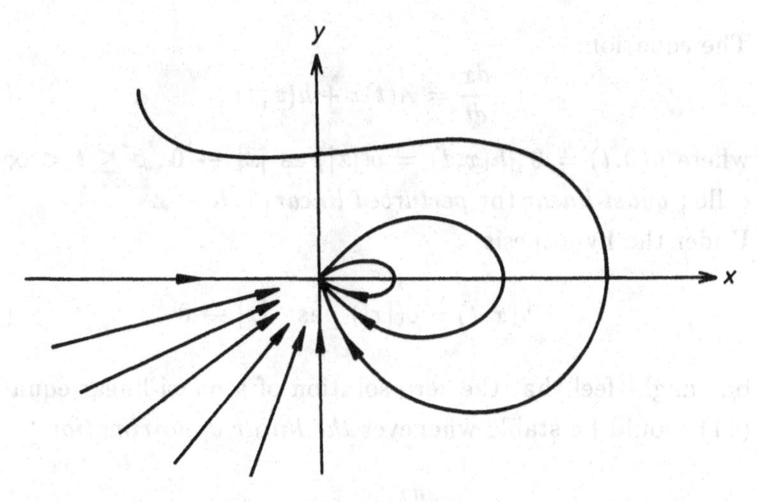

Fig. 6.2

6.3 *A nonlinear equation with unstable the zero solution and stable the linear approximation.*

Preliminaries (**76**, p. 295, 315–317):

(i) In elementary physics, the equation of quasi-linear motion

$$\theta'' + \frac{g}{l}\sin\theta = \theta'' + \frac{g}{l}(\theta - \frac{\theta^3}{3!} + \ldots) = 0 \qquad (9)$$

of the simple pendulum is often approximated by the linear equation

$$\theta'' + \frac{g}{l}\theta = 0 \qquad (10)$$

under the stipulation that $|\theta|$ be sufficiently small. This is *an example of a quasi-linear system which is approximated by its linear part.*

(ii) The trajectory of a constant solution of (**) is a critical point. *The stability of a constant solution* is thus equivalent to *the stability of its path.* The unified concept is then called *the stability of equilibrium.*

(iii) The equation

$$\frac{dx}{dt} = A(t)x + h(x,t) \ , \tag{11}$$

where $h(0,t) = 0$, $h(x,t) = o(|x|)$ as $|x| \to 0$, $\beta \le t < \infty$, is called *quasi-linear* (or *perturbed linear*) if $h = 0$.

(iv) Under the hypothesis

$$h(x,t) = o(|x|) \quad \text{as} \quad |x| \to 0 \tag{12}$$

one might feel that the zero solution of a quasi-linear equation (11) should be stable whenever *the linear approximation*

$$\frac{dx}{dt} = A(t)x \tag{13}$$

is stable. This is *not* generally true.

Example:

Consider the non-linear system

$$\left.\begin{aligned} \frac{dx}{dt} &= -y - x(x^2 + y^2) \\ \frac{dy}{dt} &= x - y(x^2 + y^2) \end{aligned}\right\} \tag{11)'$$

and the linear approximation

$$\left.\begin{aligned} \frac{dx}{dt} &= -y \\ \frac{dy}{dt} &= x \end{aligned}\right\} . \tag{13)'$$

Then the origin is a *center* for (13)′ and is *stable*, because (13)′ is of the form

$$\frac{d}{dt}\begin{bmatrix} x \\ y \end{bmatrix} = \begin{bmatrix} 0 & -1 \\ 1 & 0 \end{bmatrix} \begin{bmatrix} x \\ y \end{bmatrix} \tag{13)''$$

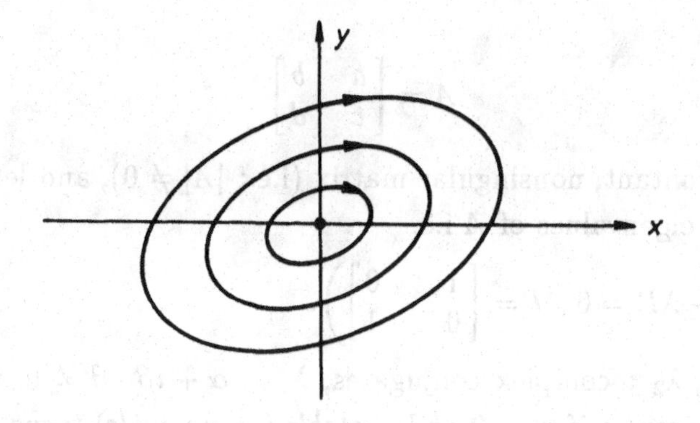

Fig. 6.3

and the matrix

$$A = \begin{bmatrix} 0 & -1 \\ 1 & 0 \end{bmatrix} \tag{14}$$

is non singular matrix (i.e.: $|A| \neq 0$), and λ_1 and λ_2, the eigenvalues of A, are:

$$|A - \lambda I| = \lambda^2 + 1 = 0, \tag{15}$$

$\lambda_1 = i\,, \lambda_2 = -i.$

The origin, however, is an *unstable focus* for (11)$'$. In fact, the solution paths of (11)$'$ are described in the polar coordinates by

$$\left.\begin{aligned} \frac{d}{dt}(r^2) &= -2r^4 \\ \frac{d}{dt}(\theta) &= 1 \end{aligned}\right\} . \tag{16}$$

Thus

$$\left.\begin{aligned} r^2 &= \frac{r_0^2}{1 + 2tr_0} \\ \theta &= \theta_0 + t \end{aligned}\right\} . \tag{16$'$}$$

Therefore the origin is *an unstable focus.*

Theorem 1: Consider the system

$$\frac{d}{dt}\begin{bmatrix} x \\ y \end{bmatrix} = \begin{bmatrix} a & b \\ c & d \end{bmatrix}\begin{bmatrix} x \\ y \end{bmatrix}, \tag{13$'''$}$$

where

$$A = \begin{bmatrix} a & b \\ c & d \end{bmatrix} \tag{14$'$}$$

is a real, contant, nonsingular matrix (i.e.: $|A| \neq 0$), and let λ_1 and λ_2 denote eigenvalues of A i.e.

$$\left(|A - \lambda I| = 0\,,\ I = \begin{bmatrix} 1 & 0 \\ 0 & 1 \end{bmatrix} \right).$$

If $\lambda_1\,,\,\lambda_2$: complex conjugates, $\lambda_1 = \alpha + i\beta\,,\,\beta \neq 0$, then the origin is a *center* if $\alpha = 0$ and a *stable* (or *unstable*) focus if $\alpha < 0$ (or $\alpha > 0$).

Theorem 2: The linear system (13) when $A: =$ constant is stable iff every eigenvalue of A has nonpositive real part, and A has m linearly independent eigenvectors corresponding to each eigenvalue with zero real part and multiplicity m.

6.4 *A quasi-linear system with unstable equilibrium and asymptotically stable linear approximation.*

The answer is given by Perron (**76**, p. 315–317).

His following example says in addition that the asymptotic stability of linear approximation does *not* imply that the zero solution of its quasi-linear equation is stable.

In fact, consider the quasi-linear system

$$\left.\begin{aligned} \frac{dx}{dt} &= -ax \\ \frac{dy}{dt} &= [\sin(\ln t) + \cos(\ln t) - 2a]y + x^2 \end{aligned}\right\}\,, \tag{17}$$

where

$$\frac{1}{2} < a < (1 + e^{-\pi})/2\,, \qquad t > 0\,. \tag{18}$$

Solving system (17) we find that all solutions are given by the formulas

$$\left.\begin{aligned} x &= x(t) = c_1 e^{-at} \\ y &= y(t) = e^{t\sin(\ln t) - 2at}\left(c_2 + c_1^2 \int_0^t e^{-s\sin(\ln s)} ds\right) \end{aligned}\right\} \tag{19}$$

We first *underestimate the integral*

$$F = F(t) = \int_0^t e^{f(s)} ds , \tag{20}$$

where

$$f = f(s) = -s\sin(\ln s) . \tag{21}$$

In fact, note first that if $0 < u < v < t$, then

$$F(t) > \int_u^v e^{f(s)} ds . \tag{22}$$

Next we find an interval (u, v) on which $f(s)$ is increasing. In fact, set

$$\sigma = \ln s . \tag{23}$$

Then

$$f'(s) = -(\sin\sigma + \cos\sigma) . \tag{24}$$

It is clear from figure 6.4 that f is increasing for

$$\frac{3\pi}{4} + 2n\pi \leq \ln s \leq \frac{7\pi}{4} + 2n\pi , \tag{25}$$

or

$$e^{\frac{3\pi}{4} + 2n\pi} \leq s \leq e^{\frac{7\pi}{4} + 2n\pi} , \quad n \geq 0 . \tag{25'}$$

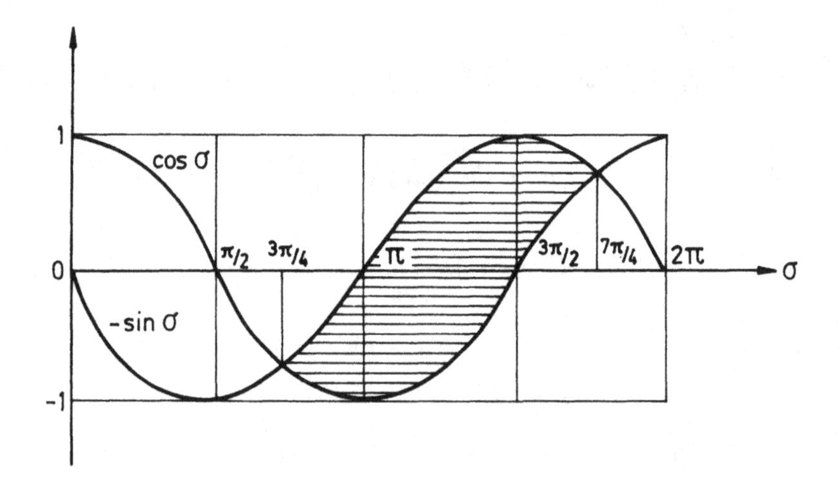

Fig. 6.4

Let

$$t_n = e^{\frac{\pi}{2}+2n\pi} \ , \ u_n = t_n e^{-\pi} \ , \ v_n = t_n e^{-\frac{3\pi}{4}} \ , \ n \geq 1 \ . \tag{26}$$

Then

$$F(t_n) \geq \int_{u_n}^{v_n} e^{f(s)} ds \geq (v_n - u_n) e^{f(u_n)} = (v_n - u_n) e^{u_n} \ .$$

Now

$$y(t_n) \geq e^{t_n \sin(\ln t_n) - 2at_n} \left\{ c_2 + c_1^2 \int_{u_n}^{v_n} e^{f(s)} ds \right\}$$

$$\geq e^{t_n(1-2a)} \left\{ -|c_2| + c_1^2 \left(e^{-\frac{3\pi}{4}} - e^{-\pi} \right) t_n \exp\left(t_n e^{-\pi} \right) \right\}$$

$$= -|c_2| e^{t_n(1-2a)} + c_1^2 \left(e^{-\frac{3\pi}{4}} - e^{-\pi} \right) t_n \exp\left(t_n (1 + e^{-\pi} - 2a) \right) \ .$$

Then, for any choice of c_2 and $c_1(\neq 0)$

$$\lim_{n\to\infty} \sup y(t_n) = \infty . \tag{27}$$

Thus *the quasi-linear system* has a critical point (0,0) which is *unstable.*

However, *the linear approximation*

$$\left.\begin{aligned} \frac{dx}{dt} &= -ax \\ \frac{dy}{dt} &= [\sin(\ln t) + \cos(\ln t) - 2a]y \end{aligned}\right\} \tag{28}$$

has a fundamental matrix solution

$$\begin{bmatrix} e^{-at} & 0 \\ 0 & e^{t\sin(\ln t)-2at} \end{bmatrix}$$

which approaches zero as $t \to \infty$.

Therefore the origin is *globally asymptotically stable for* (28).

Theorem:

(i). The linear system (13) for $t \geq \beta$ is stable iff all its solutions are bounded on intervals of the form $t_0 \leq t < \infty$, $t_0 \geq \beta$.

(ii). The linear system (13) for $t \geq \beta$ is *globally asymptotically stable* iff all its solutions approach zero as $t \to \infty$.

6.5 *The speculation that the origin would be asymptotically stable for the equation:* $\frac{dx}{dt} = f(x,t)$ *if there existed a Lyapunov function* V *with derivative* DV *negative definite is not true, in general.*

The answer is given by the following example due to Massera (**76**, p. 324–329).

Preliminaries:

(i). A real valued function $W = W(x)$ is said to be *positive definite on a neighborhood G of the origin* if

$$W = W(x) > 0 \tag{29}$$

for all $x \neq 0$ in G and $W(0) = 0$.

(ii). A function $V = V(x,t)$ defined on a cylinder $G \times [\beta, \infty)$, where G is a neighborhood of $x = 0$, is called *positive definite* if

$$V(0,t) = 0 \quad \text{for} \quad t \geq \beta \tag{30}$$

and there exists a positive definite function $W = W(x)$ on G such that

$$W(x) \leq V(x,t) \tag{31}$$

for all $(x,t) \in G \times [\beta, \infty)$.

(iii). A continuous real valued function $V = V(x,t)$ is called a *Lyapunov function for equation*

$$\frac{dx}{dt} = f(x,t) \tag{32}$$

at the origin if conditions below hold:

(1). If there is a cylinder $G \times [\beta, \infty)$, where G is the neighborhood of the origin, on which V is a positive definite, and

(2). When $\varphi = \varphi(t)$ is a solution of (32) with $\varphi(t_0) \in G$, then $V(\varphi(t), t)$ does not increase with the increasing $t \geq t_0 \geq \beta$.

(iv). If V is a continuously differentiable positive definite function on the cylinder $G \times [\beta\,, \infty)$ we define *the derivative of V with respect to equation* (32), as follows:

$$DV(x,t) = \sum_{i=1}^{n} D_i V(x,t) f_i(x,t) + \frac{\partial}{\partial t} V(x,t)\,. \tag{33}$$

Note that if $\varphi = \varphi(t)$ is a solution of equation (33), then

$$\frac{d}{dt}V(\varphi(t),t) = DV(\varphi(t),t) . \tag{34}$$

Besides it is *not* necessary to solve (32) in order to compute $DV(x,t)$. This is one of the advantages of Lyapunov's method.

(v). V is called *positive* (or *negative*) *semidefinite* if

$$V(x,t) \geq 0 \text{ (or } \geq 0) \tag{35}$$

on $G \times [\beta, \infty)$.

(vi). A continuous differentiable, positive definite function $V = V(x,t)$ is *a Lyapunov function at the origin* if DV is negative semidefinite on $G \times [\beta, \infty)$.

Example:

Consider equation

$$\frac{dx}{dt} = \frac{g'(t)}{g(t)}x , \tag{36}$$

where

$$g = g(t) = \sum_{k=1}^{\infty} \frac{1}{1 + k^4(t-k)^2} . \tag{37}$$

It is clear that g is continuous and has a continuous derivative g' given by

$$g'(t) = -\sum_{k=1}^{\infty} \frac{2k^4(t-k)}{(1 + k^4(t-k)^2)^2} , \tag{38}$$

since the indicated series converge uniformly on bounded subintervals of $\mathbb{R}^1 = (-\infty, \infty)$. It is easy to see that $g(m) > 1$ for each integer $m \geq 1$ and

$$g(t) < 2 + \sum_{k=1}^{\infty} k^{-4} = M , \ -\infty < t < \infty . \tag{39}$$

A general solution of (36) is of the form

$$x = x(t) = g(t)c \; , \tag{40}$$

where c is an arbitrary constant.

Since $g(m) > 1$ for each integer $m \geq 1$, the solution $x(t) \equiv 0$ could *not be asymptotically stable*

Define

$$V = V(x,t) = \frac{x^2}{g^2(t)} \left[M^2 + \int_t^\infty g^2(u)du \right] . \tag{41}$$

Since

$$g^2(t) < M^2 + \int_t^\infty g^2(u)du \; , \tag{42}$$

$$V(x,t) \geq x^2 \; . \tag{43}$$

Thus V is positive definite.

Moreover

$$DV(x,t) = -x^2 \; . \tag{44}$$

Therefore V is *a Lyapunov function* for (36) with DV negative definite, yet the origin is not asymptotically stable.

6.6 *A Non-linear two-dimensional real autonomous system* (**NL**) *and the corresponding linear one* (**L**) *so that a node for* (**L**) *doesn't go into the node for* (NL).

Preliminaries (**12**, p. 377):

Consider the real linear system (L):

$$\left.\begin{aligned} \frac{dx}{dt} &= ax + by \\ \frac{dy}{dt} &= cx + dy \end{aligned}\right\} \tag{45}$$

where a, b, c, d are real constants such that the determinant: $ad - bc \neq 0$. Clearly $(x, y) = (0, 0)$ is then *the only critical point of* (45), that is, the only point where the right member of (45) vanishes. System (45) can be written in the matrix form

$$\frac{d}{dt}X = AX \tag{46}$$

where

$$X = \begin{bmatrix} x \\ y \end{bmatrix}, \; A = \begin{bmatrix} a & b \\ c & d \end{bmatrix}.$$

Let A have the characteristic roots λ and μ. In the case of *a proper node* the geometry is given by the Figures 6.5–6.6, and the system (45) is given by

$$\left.\begin{aligned} \frac{dx}{dt} &= \lambda x \\ \frac{dy}{dt} &= \lambda y \end{aligned}\right\} , \lambda \neq 0 . \tag{47}$$

A solution $\varphi = \varphi(t)$ through initial point $(c_1, c_2) \neq O = (0, 0)$ is given by the formulas

$$\left.\begin{aligned} x = \varphi_1(t) &= c_1 e^{\lambda t} \\ y = \varphi_2(t) &= c_2 e^{\lambda t} \end{aligned}\right\} . \tag{48}$$

In the case of *a spiral point* the geometry is given by the Figures 6.7–6.8, and the system (45) is given by

$$\left.\begin{aligned} \frac{dx}{dt} &= \lambda x \\ \frac{dy}{dt} &= \mu y \end{aligned}\right\} , \tag{49}$$

where

$\lambda = \alpha + i\beta$, $\mu = \alpha - i\beta$ (α, β : real, $\beta \neq 0$), and either $\alpha, \beta < 0$ or $\alpha > 0$, $\beta < 0$.

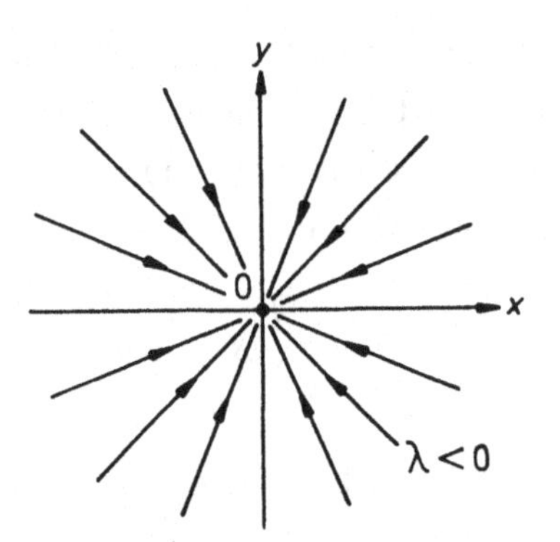

Fig. 6.5

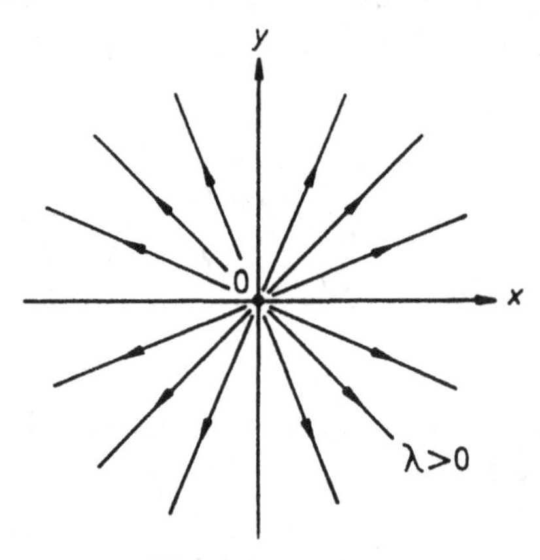

Fig. 6.6

In this case

$$\left.\begin{aligned}\frac{dx}{dt} &= \alpha x + \beta y \\ \frac{dy}{dt} &= -\beta x + \alpha y\end{aligned}\right\} \tag{50}$$

and the solution $\varphi = \varphi(t)$ passing through initial point (c_1, c_2) $(\neq O = (0,0))$, at $t = 0$, is given by the formulas

$$\left.\begin{aligned} x = \varphi_1(t) &= e^{\alpha t}(c_1 \cos\beta t + c_2 \sin\beta t) \\ y = \varphi_2(t) &= e^{\alpha t}(-c_1 \sin\beta t + c_2 \cos\beta t)\end{aligned}\right\} . \tag{51}$$

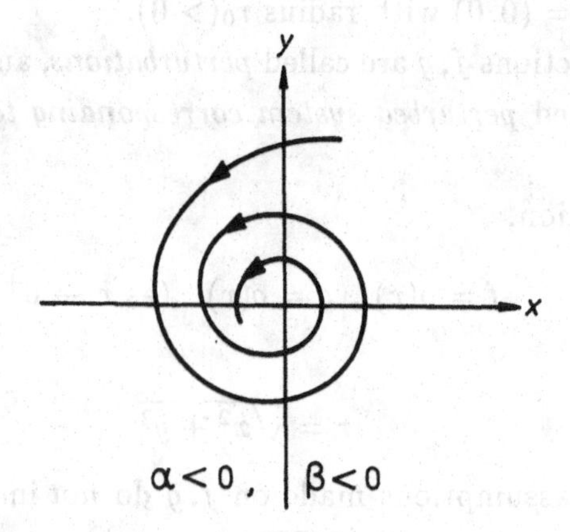

Fig. 6.7

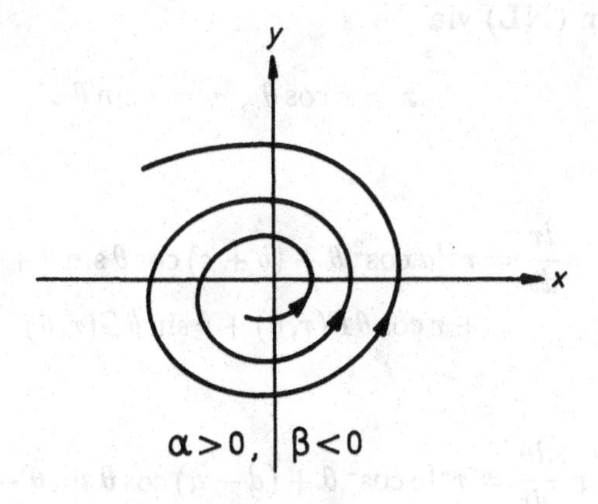

Fig. 6.8

Consider *the non-linear system* (**NL**):

$$\left.\begin{aligned} \frac{dx}{dt} &= ax + by + f(x,y) \\ \frac{dy}{dt} &= cx + dy + g(x,y) \end{aligned}\right\} \tag{52}$$

where a, b, c, d are real contants: $ad - bc \neq 0$, and $f = f(x,y)$, $g = g(x,y)$ are real continuous functions defined in some circle about the origin $(x, y) = (0, 0)$ with radius $r_0(> 0)$.

The functions f, g are called *perturbations*, and the system (**NL**) ((52)) is called *perturbed system corresponding to the linear system* (**L**)((45)).

In addition,

$$f = o(r) \ , \ g = o(r) \quad (\text{as } r \to 0^+) \ , \tag{53}$$

where

$$r = \sqrt{x^2 + y^2} \ . \tag{54}$$

Note: The assumptions made on f, g do *not* imply the uniqueness of the solutions $\varphi = \varphi(t)$ of (**NL**).

To analyze the orbits of (**NL**) is to use the *polar equations* obtained fron (**NL**) via

$$x = r\cos\theta \ , \ y = r\sin\theta \ . \tag{55}$$

Namely

$$\begin{aligned} r\frac{dr}{dt} &= r^2[a\cos^2\theta + (b+c)\cos\theta\sin\theta + d\sin^2\theta] \\ &\quad + r\cos\theta F(r,\theta) + r\sin\theta G(r,\theta) \ , \end{aligned} \tag{56}$$

and

$$\begin{aligned} r^2\frac{d\theta}{dt} &= r^2[c\cos^2\theta + (d-a)\cos\theta\sin\theta - b\sin^2\theta] \\ &\quad + r\cos\theta F(r,\theta) - r\sin\theta F(r,\theta) \ , \end{aligned} \tag{57}$$

where

$$F(r,\theta) = f(r\cos\theta, r\sin\theta) \ , \ G(r,\theta) = g(r\cos\theta, r\sin\theta) \ .$$

If $\varphi = \varphi(t) = (\varphi_1(t), \varphi_2(t))$ is a solution of (**NL**), then *the polar functions*

$$r = \rho = \rho(t) = \sqrt{\varphi_1^2(t) + \varphi_2^2(t)} \ , \ \theta = \omega = \omega(t) = \tan^{-1}\frac{\varphi_2(t)}{\varphi_1(t)} \tag{58}$$

contitute *a solution* $(\rho(t), \omega(t))$ of the polar equations (56)–(57).

If there exists a $\delta : 0 < \delta \leq r_0$, such that, for any solution path $\varphi = \varphi(t) = (\varphi_1(t), \varphi_2(t))$ of (**NL**) which has at least one point in $0 < r < \delta$, the solution exists over a t-half line, and if

$$\varphi = \varphi(t) = (\varphi_1(t), \varphi_2(t)) \to (0,0) ,$$

as $t \to \infty$ (or $-\infty$), then the origin is called *an attractor for* (**NL**).

Note: In case

$$f = g = 0 ,$$

then the nodes and spiral points are attractors, whereas saddle points and centers are *not.*

Node for (**NL**): The origin is said to be a *node* for (**NL**) if it is an attractor for which all orbits arrive at the origin in a definite direction. **Proper** *node for* (**NL**): If it is a node and every half line through the origin is tangent to some orbit there.

Spiral Point for (**NL**): The origin is said to be a *spiral point* for (**NL**) if it is an attractor such that

$$|\omega(t)| \to \infty \text{ , as } \quad t \to \infty (\text{or } t \to -\infty) ,$$

and $\varphi = \varphi(t) = (\varphi_1(t), \varphi_2(t))$ is any solution of (**NL**) whisch enters the region $0 \leq r < \delta$.

Remark for (L)

Example:

Consider the non-linear system

$$\left.\begin{aligned} \frac{dx}{dt} &= -x - \frac{y}{\ln\sqrt{x^2+y^2}} \\ \frac{dy}{dt} &= -y + \frac{x}{\ln\sqrt{x^2+y^2}} \end{aligned}\right\} \qquad (52)'$$

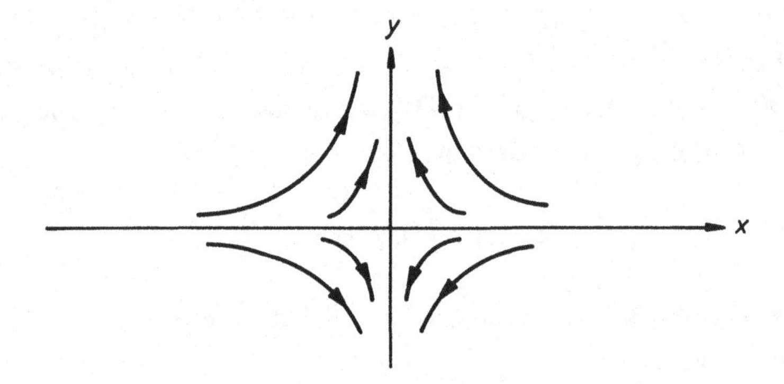

$$\frac{dx}{dt} = \lambda x\,, \frac{dy}{dt} = \mu y\text{: saddle point}(\lambda < 0 < \mu)$$

Fig. 6.9

The polar equations corresponding to (52)′ are

$$\frac{dr}{dt} = -r \tag{56$'$}$$

$$\frac{d\theta}{dt} = \frac{1}{\ln r}\,. \tag{57$'$}$$

Thus

$$r = \rho(t) = ce^{-t}\,, \ \omega = \omega(t) = -\ln(t - \ln c) + k\,, \tag{58$'$}$$

for some constant $c > 0$, and

$$k = \omega(t_0) + \ln(t_0 - \ln c)\,.$$

Therefore

$$\omega(t) \to -\infty \quad \text{as} \quad t \to \infty\,,$$

and the origin is *a spiral point for* (52)′, although the origin is *a proper node for the corresponding linear system*

$$\left.\begin{aligned} \frac{dx}{dt} &= -x \\ \frac{dy}{dt} &= -y \end{aligned}\right\}\,. \tag{45$'$}$$

6.7 *A Non-linear two-dimensional real autonomous system* (**NL**) *and the corresponding linear one* (**L**) *so that a (proper) spiral point for* (**L**) *fails to go into a (proper) spiral point for* (**NL**).

Consider the nonlinear system (**12**, **p. 378**):

$$\left.\begin{aligned}\frac{dx}{dt} &= -x + y + \frac{x}{\ln\sqrt{x^2+y^2}} \\ \frac{dy}{dt} &= -x - y + \frac{y}{\ln\sqrt{x^2+y^2}}\end{aligned}\right\} \qquad (52)''$$

The polar equation responding to (52)″, involving $\frac{dr}{dt}$, is:

$$\frac{dr}{dt}r^{-1} = -1 + \frac{1}{\ln r} \qquad (56)''$$

and this implies

$$r = \rho(t) : \rho(t)e^t = \frac{\rho_0}{\ln\rho(t) - 1}, \qquad (58)''$$

where ρ_0 is a constant.

Therefore

$$\rho(t)e^t \to 0 \quad \text{as} \quad t \to \infty$$

(for $\rho(t) \to 0$ as $t \to \infty$), and the origin is *not a (proper) spiral poibt for (52)″, although the orgin is a (proper) spiral for the corresponding linear system*

$$\left.\begin{aligned}\frac{dx}{dt} &= -x + y \\ \frac{dy}{dt} &= -x - y\end{aligned}\right\}. \qquad (45)''$$

6.8 *A Non-linear two-dimensional real autonomous system* (**NL**) *and the corresponding linear one* (**L**) *so that a center for* (**L**) *goes to a spiral point for* (**NL**).

Preliminaries: If there exists a sequence of periodic orbits $\{C_n\}$ of (NL), (52), each of which contains all later orbits and the origin in its interior and such that

$$C_n \to \quad \text{origin} \quad \text{as} \quad n \to \infty ,$$

then the origin is called *a center for* (**NL**).

Example (**12**, p. 381):

Consider the nonlinear system

$$\left.\begin{aligned} \frac{dx}{dt} &= -y - x\sqrt{x^2 + y^2} \\ \frac{dy}{dt} &= x - y\sqrt{x^2 + y^2} \end{aligned}\right\} \tag{52)'''$$

The polar equations corresponding to (52)''' are

$$\frac{dr}{dt} = -r^2 , \tag{56)'''$$

and

$$\frac{d\theta}{dt} = 1 . \tag{57)''$$

The solution of this system passing through $(r_0, \theta_0)(r_0 \neq 0)$ at $t = 0$ is given by

$$r = \rho(t) = \frac{1}{t + \frac{1}{r_0}} , \quad \theta = \omega(t) = t + \theta_0 . \tag{58)'''$$

Therefore

$$\rho(t) \to 0 \quad \text{and} \quad \omega(t) \to \infty \quad \text{as} \quad t =\to \infty ,$$

and the origin is *a spiral point for the system* (52)''', although the origin is *a center for the corresponding linear system*

$$\left.\begin{aligned} \frac{dx}{dt} &= -y \\ \frac{dy}{dt} &= x \end{aligned}\right\} . \qquad (45)'''$$

6.9 ***A Dynamical System with positively unstable motions according to Lagrange although the set Ω_p of all ω-limit points is $\neq \emptyset$.***

Preliminaries:

(i) A point q is called *an ω-limit point of a trajectory $f(p,t)$* if there exists a sequence $\{t_n\}$ such that, $t_n \to \infty$,

$$\lim_{n\to\infty} d(f(p,t_n),q) = 0 , \qquad (59)$$

where

$$d(f(p,t_n),q) := \text{distance between} \quad f(p,t_n) \quad \text{and} \quad q .$$

(ii) *A metric space R* is a set of points (elements) in which for each pair of points $p,q \in R$ there is defined a non-negative function $d(p,q)$, *distance* which satisfies axioms:

(1) $d(p,q) \geq 0$. Moreover, $d(p,q) = 0 \longleftrightarrow p = q$,

(2) $d(p,q) = d(q,p)$,

(3) $d(p,r) \leq d(p,q) + d(q,r)$ for $\forall r \in R$.

(iii) *A dynamical system* in a metric space R is a 1-parameter group $f(p,t)$, $-\infty < t < \infty, p \in R$, of transformations (mappings) of R into itself ($f(p,t) \in R$), satisfying conditions:

(1) $f(p,0) = p$,

(2) $f(p,t)$ is continuous in the pair of variables (p,t) ,

(3) $f(f(p,t_1),t_2) = f(p,t_1+t_2)$: group property.

Note: We shall call $f(p,t)$, $p :=$ fixed, *a motion*. The set of points $\{f(p,t) : -\infty < t < \infty\}, p :=$ fixed, is called *a trajectory of this motion*.

In a dynamical system there may exist motions such that all values t , $f(p,t) = p$. We shall call such a point p *a rest point* (or *point of equilibrium* or *critical point*).

(i)′ Let there be given a certain *positive half-trajectory* $f(p; 0, \infty)$. Take any bounded increasing sequence of values of t such that

$$0 \leq t_1 < t_2 < \ldots < t_n < \ldots , \lim_{n\to\infty} t_n = \infty .$$

If the sequence of points $\{f(p, t_n)\}$ has a limit point $q = \lim_{n\to\infty} f(p, t_n)$ then we shall call this point *an ω-limit point of motion* $f(p,t)$.

Note: Any limit point q' of negative half-trajectory $f(p; -\infty, 0)$ is called an *α-limit of motion* $f(p,t)$.

(iv) A motion $f(p,t)$ is called *positively stable according to Lagrange* (L^+) if the closure of the half-trajectory $f(p; 0, \infty)$ is a compact set. A motion $f(p,t)$ is called *negatively stable according to Lagrange* (L^-) if the closure of the half-trajectory $f(p; -\infty, 0)$ is a compact set.

A motion $f(p,t)$ is called *stable according to Lagrange* (L) if it is at the same time positively and negatively stable according to Lagrange.

Example (**55**, p. 340–341):

Consider the dynamical system

$$\left.\begin{array}{l} \dfrac{d\rho}{dt} = \dfrac{\rho}{1+\rho} \\ \dfrac{d\theta}{dt} = \dfrac{1}{1+\rho} \end{array}\right\} , \tag{60}$$

where $\rho \geq 0$, $-\infty < t < \infty$, and ρ, θ: polar coordinates, such that a family of motions is tracing the logarithmic spirals: $\rho = ce^{\theta}$, in the auxiliary plane $X0Y$. In fact, all the motions are continuable for $-\infty < t < \infty$. Thus we have *a dynamical system*

Moreover, all the motions are *negatively stable according to Lagrange*, having the origin (rest point) as their α-limit point. Therefore the set A_p of all α-limit points are *positively unstable according to Lagrange* since the radius vector $\rho = \rho(t) \to \infty$ as $t \to \infty$.

We now map the plane $X0Y$ on the half-plane : $-1 < x < \infty, -\infty < y < \infty$, by the transformation

$$X = \ln(1 + x)\,, \ Y = y\,. \tag{61}$$

We shall have

$$\left.\begin{aligned} \rho = \rho(t) &= (X^2 + Y^2)^{\frac{1}{2}} = (\ln^2(1+x) + y^2)^{\frac{1}{2}} \\ \theta = \theta(t) &= \arg\,(X + iY) = \tan^{-1}\left(\frac{y}{\ln(1+x)}\right) \end{aligned}\right\}. \tag{62}$$

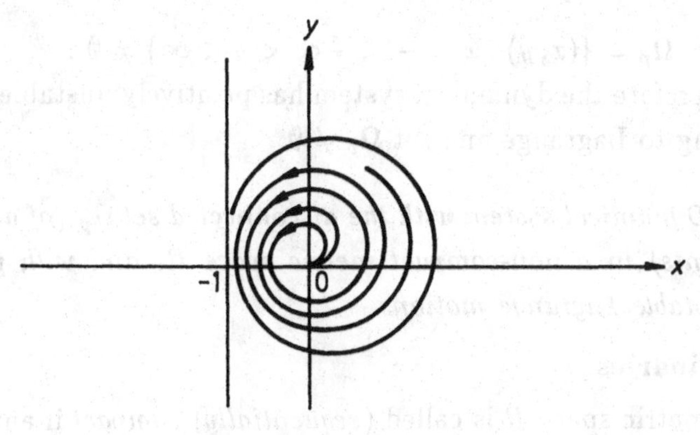

Fig. 6.10

The integral curves will have the form illustrated in Figure 6.10 and the differential equations of the new system will be

$$\left.\begin{aligned} \frac{dx}{dt} &= \frac{(1+x)[\ln(1+x) - y]}{1+\rho} \\ \frac{dy}{dt} &= \frac{y}{1+\rho} + \frac{\ln(1+x)}{1+\rho} \end{aligned}\right\}. \tag{63}$$

We complete our space with the line $x = -1$. Moreover, we shall define the the motion along it to be the limiting motion for the differential equations (as $x \to -1^+$). Since furthermore, $\frac{\ln(1+x)}{1+\rho} \to -1$, then $\frac{dy}{dt}$ remains finite and we obtainn along the line $x = -1$:

$$\left.\begin{aligned} \frac{dx}{dt} &= 0 \\ \frac{dy}{dt} &= -1 \end{aligned}\right\} . \tag{64}$$

In this way the dynamical system is defined for the closed half-plane: $x \geq -1$.

It is clear that all the motions are *positively unstable according to Lagrange* since as $t \to \infty$ they do *not* remain in a bounded part of the plane. Meanwhile, it is obvious that for any point $p = (x_0, y_0)$, $x_0 > -1$, $p \neq (0,0)$, the set Ω_p of all ω-limit points is the straight line: $x = -1$,

i.e. $\Omega_p = \{(x,y) : x = -1 \ , \ -\infty < y < \infty\} \neq \emptyset$.

Therefore the dynamical system has positively unstable motions according to Lagrange and set $\Omega_p \neq \emptyset$.

6.10 *A Dynamical system with the disconnected set Ω_p (of all ω-limit points) in a non-compact metric space R, and with positively unstable Lagrange motions.*

Preliminaries:

(i) A metric space R is called (*sequentially*) *compact* if any infinite sequence of its points contains a convergent subsequence.

(ii) A metric space R is called (*locally*) *compact* if every point $p \in R$ has a neighborhood $U(p)$ such that the closure $\overline{U(p)}$ is a compact set.

(iii) A set $E \subset R$ (: metric space) is called *connected* if it is impossible to represent it in the form

$$E = A \cup B \ ,$$

where $A, B \neq \emptyset$, and $(A \cap \overline{B}) \cup (B \cap \overline{A}) = \emptyset$.

Note: If such decomposition is possible, then the set E is called *disconnected* and A and B, if themselves connected, are called its *components.*

Examples (55, p. 343–344):

Consider the dynamical system (60).

We take the same auxiliary plane $X0Y$ (as in previous example) but this time we map the plane $X0Y$ on the strip: $-1 < x < 1, -\infty < y < \infty$, by the transformation

$$X = \frac{x}{1-x^2}\,,\ Y = y\,. \tag{61$'$}$$

We shall have

$$\left.\begin{aligned} \rho = \rho(t) &= (X^2 + Y^2)^{\frac{1}{2}} = \left(\frac{x^2}{(1-x^2)^2} + y^2\right)^{\frac{1}{2}} \\ \theta = \theta(t) &= \arg\,(X + iY) = \tan^{-1}\left(\frac{y(1-x^2)}{x}\right) \end{aligned}\right\} . \tag{62$'$}$$

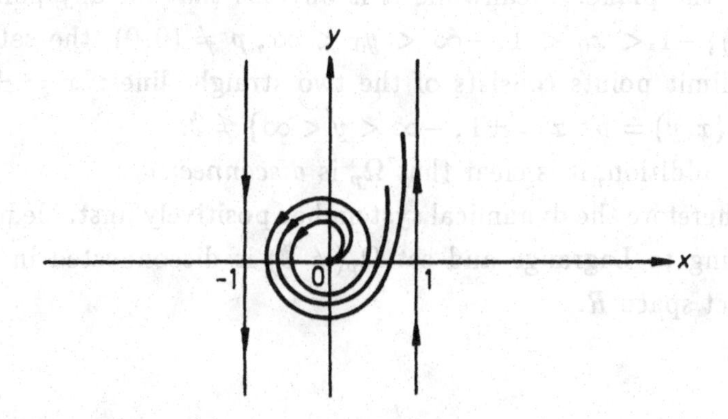

Fig. 6.11

The integral curves will have the form illustrated in Figure 6.11 and the differential equations of the new system will be

$$\left.\begin{aligned} \frac{dx}{dt} &= \frac{x(1-x^2) - y(1-x^2)^2}{(1+x^2)(1+\rho)} \\ \frac{dy}{dt} &= \frac{y}{1+\rho} + \frac{x}{1-x^2}\frac{1}{1+\rho} \end{aligned}\right\} . \tag{63)'$$

We complete our space with the lines $x = \pm 1$. Moreover, we shall define the motions along them to be the limiting motions for the corresponding limiting differential equations (as $x \to -1+$ and $x \to 1-$, with $\lim \frac{x}{(1-x^2)(1+\rho)} \to \pm 1$):

$$\left.\begin{aligned} \frac{dx}{dt} &= 0 \\ \frac{dy}{dt} &= \pm 1 \end{aligned}\right\} . \tag{64)'$$

In this way the dynamical system is defined for the closed strip: $-1 \leq x \leq 1\,,\ -\infty < y < \infty$.

It is clear that all the motions are positively *unstable according to Lagrange*, since as $t \to \infty$ they do *not* remain in a bounded part of the plane. Meanwhile it is obvious that for any point $p = (x_0, y_0)\,,\ -1 < x_0 < 1\,,\ -\infty < y_0 < \infty\,,\ p \neq (0,0)$, the set Ω_p of all ω -limit points consists of the two straight lines: $x = \pm 1$, i.e. $\Omega_p = \{(x,y) = p : x = \pm 1\,,\ -\infty < y < \infty\} \neq \emptyset$.

In addition, it is clear that Ω_p is *disconnected.*

Therefore the dynamical system has positively unstable motions according to Lagrange and set $\Omega_p(\neq \emptyset)$ is disconnected in a non-compact space R.

6.11 ***A Dynamical System with a family of solutions stable in the sense of Poincaré and Lagrange, but unstable in the sense of Lyapunov.***

Consider the dynamical system

$$\left.\begin{aligned}\frac{dx}{dt} &= -y\sqrt{x^2+y^2} \\ \frac{dy}{dt} &= x\sqrt{x^2+y^2}\end{aligned}\right\}, \tag{65}$$

which accepts the two parameter family of the solutions (**50**, p. 546–547):

$$x = a\cos(at+b)\ ,\ y = a\sin(at+b)\ , \tag{66}$$

where a and b are arbitrary.

This family of solutions (66) is stable in *Poincare* and *Lagrange sense*, but it is *unstable* in *Lyapunov sense.*

Note:

By introducing new variables, one can make the solution *Lyapunov stable.* In case the new variables r and b are introduced by means of the relations

$$\left.\begin{aligned}x = r\cos\theta,\ y = r\sin\theta \\ \theta = at+b\end{aligned}\right\}, \tag{67}$$

the original system (65) is transformed into the new system

$$\left.\begin{aligned}\frac{dr}{dt} &= 0 \\ \frac{db}{dt} &= 0\end{aligned}\right\} \tag{68}$$

for which the solutions

$$r = c_1\ ,\ b = c_2 \tag{69}$$

are Lyapunov stable, where c_1 , c_2: arbitrary contants.

6.12 *The solutions of the equation of pendulum:* $\frac{d^2x}{dt^2} + \sin x = 0$ *are, exept for the origin, unstable in Lyapunov sense, but using new coordinates are Lyapunov stable.*

Consider equation of pendulum

$$\frac{d^2x}{dt^2} + \sin x = 0 , \tag{70}$$

which accepts the two parameter family of solutions (**50**, p. 547):

$$x = a\sin(\varphi(a)t + b) , \tag{71}$$

where a, b are arbitrary, and $\varphi = \varphi(a)$ can be expressed in terms of *elliptic functions.*

The solutions (71) are, except for the origin, *unstable in Lyapunov sense.*

By introducing new variables, r and b, according to the transformation formulas

$$x = r\cos(\varphi(r)t + b) , \ y = r\sin(\varphi(r)t + b) \tag{72}$$

the system (70) leads to the new system (68) of which the solutions (69) are *Lyapunov stable.*

6.13 *A Dynamical System with motions stable according to Poisson (including the rest points) and with motions unstable according to Poisson in both directions.*

Preliminaries (**55**, p. 346–347):

(i) A point p is called *positively stable according to Poisson* (P^+) if for any neighborhood U of the point p and for any $T > 0$, there can be found a value $t \geq T$ such that $f(p,t) \in U$.

(ii) Analogously, if there can be found a $t \leq -T$ such that $f(p,t) \in U$, then the point p is *negatively stable according to Poisson* (P^-).

(iii) A point stable according to Poisson both as $t \to \infty$ and as $t \to -\infty$ is called (simply) *stable according to Poisson* (P).

(iv) The simplest example of *a motion stable* (P), *neither a rest point nor a periodic motion*, is the motion *on the surface of a torus* (Figure 6.12):

$$D_t = \{(\varphi,\theta) : 0 \leq \varphi < 1\,,\, 0 \leq \theta < 1\,, and$$
$$(\varphi + k\,,\, \theta + k') \equiv (\varphi,\theta)\,,$$
$$k, k' : \text{integers, and } \varphi, \theta :$$
$$\frac{d\varphi}{dt} = 1\,,\ \frac{d\theta}{dt} = \alpha := \text{positive irrational}\}\,.$$

Here the trajectory A of each motion is everywhere dense (i.e.: Set $A \subset R$: metric space is everywhere dense in R if $\overline{A} = R$) on the torus, every motion is stable (P), and sets Ω_p and A_p for any point $p = (\varphi, \theta)$ coincide with the surface of the torus.

Note: The *only* trajectories in the plane stable (P) are rest points and the trajectories are periodic motions.

Example:

Consider the dynamical system

$$\frac{d\varphi}{dt} = \Phi(\varphi,\theta)\,,\ \frac{d\theta}{dt} = \alpha\Phi(\varphi,\theta)\,. \tag{73}$$

Define motions on the torus by the system (73), where $\Phi = \Phi(\varphi,\theta)$ is a continuous function on the torus (periodic in the arguments φ, θ with periodic 1) satisfying the Lipschitz condition and positive everywhere except for the set of the points

$$(0,0)\,,\,(0,\theta_k)\,,\,(0,\theta'_k)\quad (k = 1,2,\ldots)$$

where it vanishes. However, α is expanded as an *infinite continued fraction* with its consecutive convergents to be written $\frac{p_k}{q_k}$,

$$\frac{p_2}{q_2} < \frac{p_4}{q_4} < \ldots < \alpha < \ldots < \frac{p_3}{q_3} < \frac{p_1}{q_1},$$

which implies that as rest points there can be taken the points with coordinates ($\varphi = 0, \theta$):

$$\theta_k \equiv \alpha q_k \ (\text{mod } 1)\ ,\ 0 < \theta_k < 1\ ,$$
$$\theta'_k \equiv -\alpha q_k \ (\text{mod } 1)\ ,\ 0 < \theta'_k < 1\ ,$$

p_k, q_k : integers.

Note:

$$|\theta_k| = |\alpha q_k - p_k| = q_k|\alpha - \frac{p_k}{q_k}| < q_k \frac{1}{q_k^2} = \frac{1}{q_k}$$

and analogously for θ'_k.

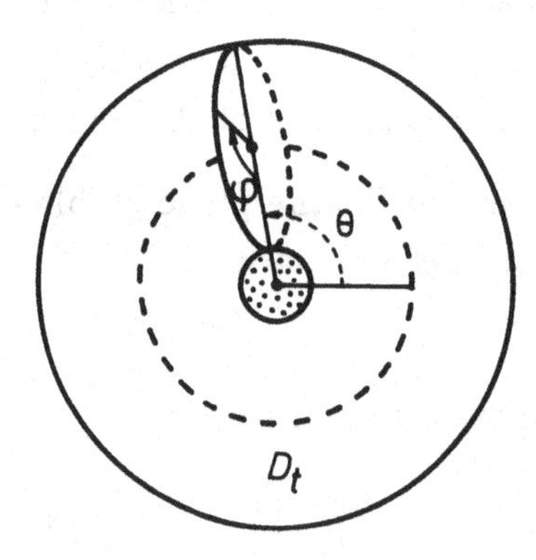

Fig. 6.12

The corresponding equations of the form (73) will possess the same trajectories as the equations

$$\frac{d\varphi}{dt} = 1\ ,\ \frac{d\theta}{dt} = \alpha \tag{73'}$$

so they will be *stable* (P) *except* for the trajectories lying on the curve

$$\theta = \alpha\varphi \,. \tag{74}$$

This latter breaks up into the countable set of arcs

$$0 < \varphi < q_1 \,, \quad q_1 < \varphi < q_2 \,, \dots \,,$$
$$0 > \varphi > -q_1 \,, \quad -q_1 > \varphi > -q_2 \,, \dots$$

separated by the rest points.

Along each such arc, for example,

$$q_k < \varphi < q_{k+1} \,,$$

the motion will be *unstable* (P) (in both directions) since as $t \to \infty$ it approaches the rest point

$$\left.\begin{array}{l} \varphi = q_{k+1} \equiv 0 \pmod 1 \\ \theta \equiv \alpha q_{k+1} \pmod 1 \end{array}\right\} \tag{75}$$

and analogously as $t \to -\infty$ it approaches the point

$$\left.\begin{array}{l} \varphi = q_k \equiv 0 \pmod 1 \\ \theta \equiv \alpha q_k \pmod 1 \end{array}\right\} . \tag{76}$$

6.14 *An ill-conditioned initial-value problem*

Preliminaries: Assume the initial-value problem

$$\left.\begin{array}{l} \dfrac{dy}{dx} = f(x, y) \,, \quad x_0 - a \leq x \leq x_0 + a \\ y(x_0) = y_0 \end{array}\right\} . \tag{77}$$

Solving (77) numerically, we will generally assume that the solution $y = \varphi(x)$ is being sought on a given finite interval: $x_0 \leq x \leq b$.

In that case, it is possible to obtain the following *stability result*:

Make a small change in the initial value for the initial-value problem, changing y_0 to $y_0 + \epsilon$. Call the resulting solution $y_\epsilon(x)$:

$$\left.\begin{aligned} &\frac{dy_\epsilon}{dx} = f(x, y_\epsilon) \ , \ x_0 \le x \le b \\ &y_\epsilon(x_0) = y_0 + \epsilon \end{aligned}\right\} . \tag{77)'}$$

Let $f(x, y_\epsilon)$ and $\frac{\partial f}{\partial y_\epsilon}(x, y_\epsilon)$ be continuous functions of x and y_ϵ at all points (x, y_ϵ) in some neighborhood of the initial point $(x_0, y_\epsilon(x_0))$. Then it can be shown that for all small values of ϵ,

$$E = \max_{x_0 \le x \le b} |y_\epsilon(x) - y(x)| \le c\epsilon \tag{78}$$

for some $c > 0$. Problem (77)′ is called *perturbed.*

Thus *small changes in the initial value* y_0 *will lead to small changes in the solution* $y = \varphi(x)$ *of the problem* (77).

If the *maximum error* E is much larger than ϵ, then (77) is said to be *ill-conditioned.*

Example (1, p. 288–292):

Consider the initial-value problem

$$\left.\begin{aligned} &\frac{dy}{dx} = \lambda(y - 1) \ , \ 0 \le x \le b \\ &y(0) = 1 \end{aligned}\right\} \tag{79}$$

where $\lambda < 0$. The perturbed problem

$$\left.\begin{aligned} &\frac{dy_\epsilon}{dx} = \lambda(y_\epsilon - 1) \ , \ 0 \le x \le b \\ &y_\epsilon(0) = 1 + \epsilon \end{aligned}\right\} \tag{79)'}$$

has the solution

$$y_\epsilon = 1 + \epsilon\, e^{\lambda x} \ , \ x \ge 0 \ . \tag{80}$$

For the *error*

$$y - y_\epsilon = y(x) - y_\epsilon(x) = \epsilon e^{\lambda x} \tag{81}$$

(where $y = 1$: the solution of (79)),

$$E = \max_{0 \le x \le b} |y_\epsilon - y| = |\epsilon| \, , \; \lambda < 0 \, . \tag{82}$$

The error (81) decreases as x increases.

Therefore problem (79) is ill-conditioned.

6.15 *The Cauchy problem for Laplace equation:* $\Delta u = u_{xx} + u_{yy} = 0$ *with data* $u(x,o) = \tau(x)$, $u_y(x,0) = \nu(x)$, $0 \le x \le 1$, *is unstable in Hadamard sense in the metric of class* $\mathbf{C}^1$ *(or any space* $\mathbf{C}^m$, $m = 1, 2, \ldots$, *or even any space* $\mathbf{L}^2_m$*).*

Definition: A mathematical problem which is to correspond to physical reality is *a properly posed* (or *a well-posed*) *problem in Hadamard sense* if it satisfies the following three requirements:

(i) The solution must exist.

(ii) The solution should be uniquely determined.

(iii) The solution should be stable (that is, to depend continuously on the data).

If condition (iii) breaks down (fails to hold), then the mathematical problem is *unstable in Hadamard sense.*

A mathematical problem cannot be considered as realistically corresponding to physical phenomenon unless *a variation of the given data in a sufficiently small range leads to an arbitrary small change in the solution.* This requirement of "*stability in Hadamard sense*" is not only essential in mathematical physics, but also in approximation theory.

Example (**42**, p. 125–126; **23**, p. 108):

Consider the Cauchy problem

$$\left.\begin{array}{r}\Delta u \equiv u_{xx} + u_{yy} = 0 \\ u(x,0) = \tau(x) \ , \ \ u_y(x,0) = \nu(x)\end{array}\right\} , \tag{83}$$

$0 \leq x \leq 1$, where

$$\tau(x) = 0 \ , \ \nu(x) = \frac{\pi}{n} \sin(\pi n x) \ . \tag{84}$$

The functions

$$u^{(n)} = u^{(n)}(x,y) = \frac{1}{n^2} \sin(\pi n x) \sin h(\pi n y) \ , \tag{85}$$

$$n = 1, 2, \ldots \ , \ \ 0 \leq x \leq 1 \ , \ \ 0 \leq y \leq 1 \ ,$$

satisfy Problem (83)–(84).

It is evident that the sequences

$$\left\{\tau^{(n)}(x)\right\} \ , \ \left\{(\tau^{(n)}(x))'\right\} \ , and \ \left\{\nu^{(n)}\right\}$$

tend to zero uniformly in the interval $[0,1]$, such that

$$u^{(n)}(x,0) = \tau^{(n)}(x) = 0 \ , \ \frac{\partial u^{(n)}}{\partial y}(x,0) = \nu^{(n)}(x) = \frac{\pi}{n} \sin(\pi n x) \ . \tag{86}$$

However for an arbitrary n there exists a number $x_n \in (0,1)$, such that $\sin(\pi n x_n) = 1$ which implies that

$$\lim_{n \to \infty} \sup \left\{ |u^{(n)}(x,y)| : 0 \leq x \leq 1 \ , \ 0 \leq y \leq 1 \right\} = \infty \ .$$

This proves that the solution of the Cauchy problem for equation (83) does *not* depend continuously (in the metric $\mathbf{C}^1$) on the initial data (84) given for $y = 0$ in the interval $0 \leq x \leq 1$.

This example is due to J. Hadamard.

6.16 *The Cauchy problem:* $u_{tt} + u_x = 0$, $u(x,0) = 0$, $u_t(x,0) = \frac{1}{n}\exp(-n^2 x)$ *is unstable in Hadamard sense.*

Consider equation (14, p. 486):

$$u_{tt} + u_x = 0 \tag{87}$$

with data

$$u(x,0) = 0 \ , \ u_t(x,0) = \frac{1}{n} e^{-n^2 x} \ . \tag{88}$$

The solution of problem (87)–(88) is of the form

$$u = \frac{1}{n^2} \frac{e^{nt} - e^{-nt}}{2} e^{-n^2 x} \ . \tag{89}$$

It is clear that, if $n \to \infty$, we have for $t = 0$ uniformly $u = 0$, $u_t \to 0$ while $u = u(x,t)$ diverges in every region for $t > 0$.

The initial value problem for vanishing initial values of u and u_t has identically vanishing solution.

Thus the continuity stability condition is violated and Cauchy problem is unstable in Hadamard sense.

Remark:

G. I. Taylor has shown that an important question of *stability* leads to *a meaningful improperly posed problem*:

Consider a system of two incompressible fluids separated by an interface, moving toward the lighter of the fluids. This phenomenon can be described by means of a velocity potential. This potential turns out to be a solution of an improper initial value problem for Laplace equation.

Overdetermined problems form another type of meaningful "improper" problem. For instance, we may seek a function, harmonic inside the unit circle, which has prescribed values in the concentric circle of radius $\frac{1}{2}$.

7. SINGULARITIES

(For *singular points* of systems in "Stability Section")

7.1 *A Differential Equation with singular locus containing no singular solutions.*

Definition: The locus of all points (x,y), such that for some $y'(=\frac{dy}{dx}=p)$

$$F(x,y,y')=0 \ , \ \frac{\partial F}{\partial y'}(x,y,y')=0 \tag{1}$$

hold, is called *singular locus*, or the p - *discriminant locus.*

The singular locus may contain a solution $y=\varphi(x)$ of equation

$$F(x,y,y')=0 \ . \tag{2}$$

Such a solution is called *a singular solution of* (2).

Example (**36**, p. 212–213):

Consider the differential equation

$$F(x,y,y')\equiv x^2+y^2+(y')^2-1=0 \ . \tag{2$'$}$$

The surface (2)$'$ in xyy'-space is a sphere, whose tangent plane at $P=(x_0,y_0,y_0')$:

$$\left.\frac{\partial F}{\partial x}\right|_P (x-x_0)+\left.\frac{\partial F}{\partial y}\right|_P (y-y_0)+\left.\frac{\partial F}{\partial p}\right|_P (p-p_0)=0 \tag{3}$$

is vertical where

$$x^2+y^2=1 \ . \tag{4}$$

In fact, *the exceptional points* (4) are precisely those at which (1) hold. That is,

$$\frac{\partial F}{\partial y'}=2y'=0 \ , \ \text{or } y'=0 \ .$$

Therefore *the singular locus* is the curve (4) which can be represented by two functions of x:

$$y = \pm\sqrt{1 - x^2}\ . \tag{5}$$

One or both of these may be singular solutions. Formula (5) yields (by differentiation):

$$y' = \pm\frac{x}{\sqrt{1 - x^2}}\ . \tag{6}$$

Thus from (5), (2)′ and (6) we obtain

$$\begin{aligned} x^2 + y^2 + (y')^2 - 1 &= x^2 + (1 - x^2) + \frac{x^2}{1 - x^2} - 1 \\ &= \frac{x^2}{1 - x^2} \neq 0\ ,\ x \neq 0\ . \end{aligned} \tag{7}$$

Hence the differential equation (2)′ is satisfied only at $x = 0$. Therefore there are no singular solutions.

7.2 *A class of differential equations of 2nd order with no singular solutions.*

Definition: For a second order equation

$$F(x, y, y', y'') = 0 \tag{8}$$

we reduce it to first degree by solving for y''. This is possible except where

$$\frac{\partial F}{\partial y''}(x, y, y', y'') = 0\ . \tag{9}$$

A solution $y = \varphi(x)$ of equation (8) such that (9) holds at each point is called *a singular solution.*

Note: Similarly, for an nth order equation

$$F(x, y, y', y'', \ldots, y^{(n)}) = 0 \tag{10}$$

a singular solution is a solution $y = \varphi(x)$ such that

$$\frac{\partial F}{\partial y^{(n)}}(x, y, y', y'', \ldots, y^{(n)}) = 0 \tag{11}$$

holds at each point.

Example (**36**, p. 216)

All linear equations

$$a_0(x)y'' + a_1(x)y' + a_2(x)y = f(x) , \tag{12}$$

where $a_0(x) \neq 0$, have *no* singular solutions.

7.3 *A differential equation of 1st order with an envelope of the family of solutions which is not a singular solution*

Definition: Consider the family of curves

$$G(x, y, c) = 0 \tag{13}$$

in implicit form, for each constant c, such that (13) represents one or more curves in the xy-plane.

A curve in the xy-plane is called *an envelope of the family* (13) if at each point of this curve at least one member of the family is tangent to the curve.

The locus in the xy-plane described by the equations

$$G(x, y, c) = 0 , \ \frac{\partial G}{\partial c}(x, y, c) = 0 \tag{14}$$

is called c - *discriminant locus.*

Example (**36**, p. 215):

Consider equation

$$y' - y^{\frac{2}{3}} = 0 . \tag{15}$$

It is obvious that the line

$$y = 0 \tag{16}$$

is an *envelope* of the family of solutions of (15) but is *not* a singular solution of (15).
Note: $F = y' - y^{\frac{2}{3}}$, $F_y = -\frac{2}{3}y^{-\frac{1}{3}}$. Thus *no* uniqueness on $y = 0$.

7.4 *A differential equation of the 1st order with a singular solution which is not an envelope of the remaining solutions.*

Consider equation (**36**, p. 215):

$$(y')^3 - y^3 = 0\ . \tag{17}$$

It is clear that the line (16) is a singular of solution (17) but *not* an envelope of the remaining solutions.

7.5 *A homogeneous linear differential equation of 2nd order with no solutions of the form* $\sum_{n=0}^{\infty} c_n x^n$ *(except* $y = 0$*).*

Consider equation (**36**, p. 235)

$$x^2 y'' + xy' + y = 0\ . \tag{18}$$

This is *Cauchy-Euler equation* of 2nd order and is a special case of equation

$$a_0 x^n y^{(n)} + a_1 x^{n-1} y^{(n-1)} + \ldots + a_{n-1} x y' + a_n y = 0\ . \tag{19}$$

It is clear that

$$y = x^r \tag{20}$$

is a solution of the related homogeneous equation

$$\begin{aligned} a_0 r(r-1)\ldots(r-n+1) + a_1 r(r-1)\ldots(r-n+2) \\ + \ldots + a_{n-1} r + a_n = 0\ . \end{aligned} \tag{21}$$

If (21) has distinct real roots: $r_1, r_2, \ldots, r_n$, we obtain easily the general solution of (19).

In (18) we have

$$a_0 = a_1 = a_2 = 1 \ , \ n = 2 \tag{22}$$

and thus (21) takes the form

$$r(r-1) + r + 1 = 0 \ , \tag{23}$$

or

$$r^2 + 1 = 0 \ , \tag{23'}$$

which has *no* real roots.

Note: The substitution

$$x = e^t \tag{24}$$

reduces equation (19) to one with constant coefficients for y in terms of t.

7.6 Implicit Function Theorem:

Let $F(x, y, y')$ be continuous and have continuous partial derivatives with respsct to x, y and y' in some region D of x, y, y'- space.

If $P = (x_0, y_0, y_0')$ is a triple D such that

$$F(P) = F(x_0, y_0, y_0') = 0 \ , \tag{25}$$

and $(p = y', p_0 = y_0')$

$$\frac{\partial F}{\partial p}(x_0, y_0, p_0) \neq 0 \ , \tag{26}$$

then there is a unique function

$$p(= y') = f(x, y) \tag{27}$$

defined for x, y close enough to x_0, y_0, respctively, such that

$$p_0(= y_0') = f(x_0, y_0) \tag{27'}$$

and satisfying equation (2).

Note: Employing **I. F. T** (Implicit Function Theorem) an equation of the form (2) can be reduced to a 1st degree form by solving for y'.

For instance, equation

$$(y')^2 + y'(x^2y - xy) - x^3y^2 = 0 \tag{28}$$

or equation

$$(y' - xy)(y' + x^2) = 0 \tag{28'}$$

is satisfied at

$$P : x = 0\,,\ y = 0\,,\ y' = 0\,. \tag{29}$$

However,

$$\frac{\partial F}{\partial y'} = 2y' = 0 \quad \text{at} \quad P$$

and no function $y' = f(x, y)$ is obtained.

Therefore the above equation (28) is *an example of an equation of the form (2) which is not reduced to a 1st degree form:* $y' = f(x, y)$.

A similar *example* is equation

$$(y')^2 - 4y = 0\,, \tag{30}$$

which is not satisfied for any values of x, y, y' (in this case *no* function: $y' = f(x, y)$ is obtained).

8. DYNAMICAL SYSTEMS

(For more about Dynamical systems in "Stability Section")

8.1 *A Dynamical System whose integral curves recede in both directions but whose families of trajectories are , nevertheless, not regular.*

Definition (due to Hassler Whitney):

Consider a family S of integral curves filling either a region G or a closed region $\overline{G}$ in $\mathbb{R}^n$. A family S of trajectories filling a domain G (*not* necessarily open) in $\mathbb{R}^n$, is called *a regular family* if there exists a homeomorphism (one to one and bicontinuous mapping) of the domain G onto the set $E \subset \mathbb{R}^n$ or $\mathbb{R}^{n+1}$, which maps trajectories into parallel straight lines so that the images of different integral curves lie on different straight lines.

Example (**55**, p. 30–31):

Consider the dynamical system

$$\left.\begin{aligned} \frac{dx}{dt} &= \sin y \\ \frac{dy}{dt} &= \cos^2 y \end{aligned}\right\} . \tag{1}$$

The integral curves of this system are

$$\frac{1}{\cos y} = x + c \tag{2}$$

and the straight lines

$$y = k\pi + \frac{\pi}{2} \; , \; k = 0, \pm 1, \ldots \; . \tag{3}$$

We consider only the strip

$$\overline{G} : \quad -\frac{\pi}{2} \le y \le \frac{\pi}{2} \; , \; -\infty < x < \infty \; .$$

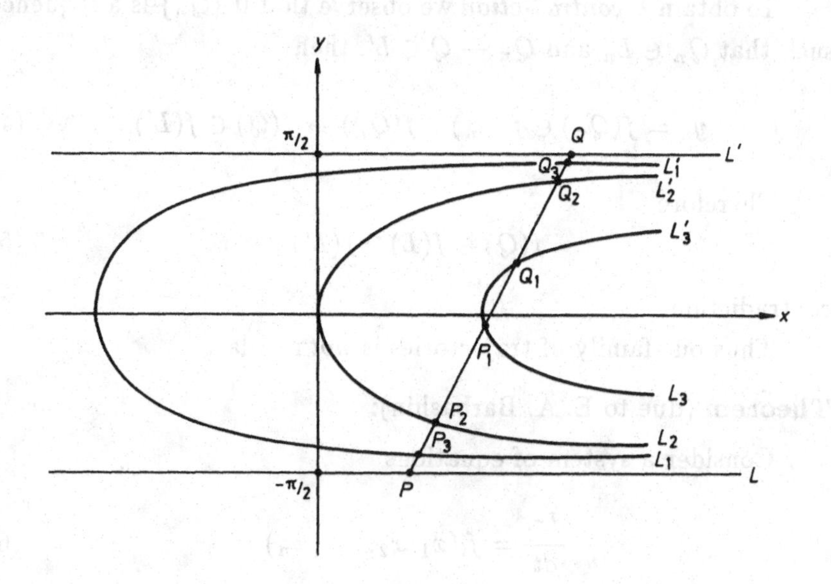

Fig. 8.1

Although all the integral curves situated within the above strip recede in both directions, the family of integral curves filling this strip is not regular.

To prove this, draw a segment PQ with the endpoints P, Q on the lines $Y = -\frac{\pi}{2}$, $y = \frac{\pi}{2}$, respectively, and consider a sequence of points P_n on this segment, converging to P.

Write L_n for the trajectory passing through P_n, and L and L' respectively for the lower and upper boundaries of the strip.

Assume that our family of trajectories is *regular*, and let f be a homeomorphism (of above definition).

Then the sequence $f(P_n) \in f(L_n)$ converges to the point $f(P) \in f(L)$.

Moreover any convergent sequence of points $y_n \in f(L_n)$ has its limit point of $f(L)$, since $f(L_n)$ and $f(L)$ are parallel straight lines.

To obtain *a contradiction* we observe that if $\{Q_n\}$ is a sequence such that $Q_n \in L_n$ and $Q_n \to Q \in L'$, then

$$y_n = f(Q_n) \in f(L_n) \ , \ f(Q_n) \to f(Q) \in f(L') \ . \qquad (4)$$

Therefore

$$f(Q) \in f(L) \cap f(L') \ , \qquad (5)$$

contradiction.

Thus our family of trajectories is *not* regular.

Theorem (due to E. A. Barbashin):

Consider a system of equations

$$\frac{dx_i}{dt} = f_i(x_1, x_2, \ldots, x_n) \ , \qquad (6)$$

$i = 1, 2, \ldots, n$, which defines a dynamical system in a domain G. If there exists a single-valued function $u = u(x_1, x_2, \ldots, x_n)$ satisfying condition

$$\sum_{i=1}^{n} \frac{\partial u}{\partial x_i} f_i = 1 \qquad (7)$$

in G, then our dynamical system is *regular*.

Note: The conclusions of this Theorem still hold if we replace condition (7) by

$$N = \sum_{i=1}^{n} \frac{\partial u}{\partial x_i} f_i \geq K^2 \ (> 0) \ . \qquad (8)$$

8.2 *Nonperiodic solutions stable in the sense of Poisson*

Introduce real Cartesian coordinates (φ, θ) in the plane (**55**, p. 27–29).

Identify any two points

$$(\varphi,\theta)\ ,\ (\varphi+n\,,\theta+m)$$

whose coordinates differ by integers n, m respectively.

On the resulting *torus* (Figure 6.12):

$$\left.\begin{aligned}\frac{d\varphi}{dt}&=1\\ \frac{d\theta}{dt}&=\alpha\end{aligned}\right\}, \tag{9}$$

there are two different cases:

I. $\alpha = \frac{p}{q}$ (rational number).

II. $\alpha :=$ irrational.

Case I. Consider the integral curves of equation

$$\frac{d\theta}{d\varphi}=\frac{p}{q}\ , \tag{10}$$

where q: natural number, p: integer, and $\frac{p}{q}$: irreducible. The solution corresponding to the initial conditions

$$\varphi = 0\ ,\ \theta=\theta_0 \tag{11}$$

has the form

$$\theta=\theta_0+\frac{p}{q}\varphi\ . \tag{12}$$

As φ takes on the value q, the coordinate θ in (12) takes the value θ_0+p, the resulting point of our integral curve on the torus coincides with the initial point $(0,\theta_0)$ and the curve is closed.

Thus the torus is covered by closed integral curves of (10).

Case II. Consider next equation

$$\frac{d\theta}{d\varphi}=\alpha\ (:\text{irrational})\ . \tag{13}$$

In this case there are *no* closed curves (thus *non-periodic solutions*) among the integral curves

$$\theta = \theta_0 + \alpha\varphi \tag{14}$$

of (13). In fact, suppose that a point (φ_1, θ_1) on the integral curve (14) coincides with the initial point $(0, \theta_0)$, then

$$\theta_1 = \theta_0 + \alpha\varphi_1 = \theta_0 + \alpha n = \theta_0 + m$$

$(m, n$: integers) whence

$$\alpha n = m \quad \text{and} \quad \alpha = \frac{m}{n} : \text{rational} ,$$

which is a *contradiction.*

Since all the trajectories can be obtained from the trajectory

$$\theta = \alpha\varphi \tag{14$'$}$$

by a translation along the θ axis, we need to consider only this trajectory in detail. Its intersections with the meridian

$$\varphi = 0 \tag{15}$$

are

$$\varphi = 0 \ , \ \theta_n = \alpha n \ , \ n = 0, \pm 1, \ldots \ . \tag{16}$$

These points are *everywhere dense in this meridian* (A set $A \subset R$: metric space is called everywhere dense in R if $\overline{A} = R$).

In fact, write

$$(\alpha) = \alpha - [\alpha] \ , \tag{17}$$

where $[\alpha]$: = greatest integer not greater than α. It is enough to show that the set

$$\{(\alpha n) \ , \ n = 0, \pm 1, \ldots\} \tag{18}$$

is *everywhere dense in the interval* $[0,1]$.

Indeed, since α: = irrational, the $p+1$ numbers:

$$0 , (\alpha) , \dots , (\alpha p) \tag{19}$$

are all distinct and since they are all distributed among the p intervals

$$I_h : \frac{h}{p} \leq \theta \leq \frac{h+1}{p} , \; h = 0,1,\dots ,p-1 , \tag{20}$$

one of this intervals must contain at least two of the numbers (19). Let $(\alpha k_1) , (\alpha k_2)$ be two such numbers.

They differ by less than $\frac{1}{p}$ since each of I_h is of length $\frac{1}{p}$.

If $k_2 > k_1$, we write

$$k = k_2 - k_1 \, . \tag{21}$$

Then either

$$(\alpha k) \in I_0 \quad \text{or} \quad (\alpha k) \in I_{p-1} \, . \tag{22}$$

In either case, the sequence

$$(\alpha \cdot k) , (\alpha \cdot 2k) , (\alpha \cdot 3k) , \dots \tag{23}$$

continued as long as may be necessary, will partition the interval $[0,1]$ into segments of less than $\frac{1}{p}$.

Besides everywhere ϵ-neighborhood of a point in $[0,1]$, contains a point of the set $\{(\alpha n)\}$ if we take

$$p > \frac{1}{\epsilon} \tag{24}$$

in the above discussion.

Thus the set $\{(\alpha n)\}$ *is everywhere dense in* $[0,1]$. Therefore every point of the meridian (15) is a limit point for the set

$$\varphi = n \ , \ \theta = \alpha n \tag{25}$$

of our trajectory. Similarly, every point

$$\varphi = \varphi_0 \ , \ \theta = \theta_0 \tag{26}$$

is the limit point for the set of points

$$\varphi = \varphi_0 + n \ , \ \theta = \alpha(\varphi_0 + n) \tag{27}$$

of the above trajectory.

It follows that trajectory $(14)'$ and hence every trajectory of (13) is everywhere dense in the torus.

In particular, every trajectory, even though it is *not* closed, contains some of its ω- limit points.

The fact that every trajectory is *not* closed means that we have *non-periodic solutions*, and that it contains some of its ω- limit points means that we have *Poisson stability.*

Therefore we have non-periodic solutions stable in Poisson sense on the torus.

8.3 *A Dynamical System With Non-uniform Motion*

Define motions on the torus (Fig. 6.12) by the equations (**55**, p. 346):

$$\left.\begin{aligned} \frac{d\varphi}{dt} &= \Phi(\varphi,\theta) \\ \frac{d\theta}{dt} &= \alpha\Phi(\varphi,\theta) \end{aligned}\right\} , \tag{28}$$

where $\Phi = \Phi(\varphi,\theta)$ is a continuous function on the torus (periodic in the arguments φ, θ with period 1) everywhere positive except at the point (0,0), where $\Phi(0,0) = 0$, and satisfying Lipschitz condition.

The curves along which the motions take place remain the same as in (9) since thay are determined by

$$\frac{d\theta}{\alpha} = \frac{d\varphi}{1}, \tag{29}$$

but the character of the motion has been altered.

Along the curve (14)′ *there are three kinds of motion*:

(i) $\theta = 0$, $\varphi = 0$ (i.e. rest point).

(ii) Motions along the positive arc $0 < \varphi < \infty$. For this kind of motion the positive half-trajectory is everywhere dense on the torus and therefore *is stable* (P^+) (i.e. positively stable in the Poisson sense).
The negative half-trajectory tends to the rest point (0,0) as $t \to -\infty$ and thus it is *unstable* P^- (i.e. negatively unstable in the Poisson sense).

(iii) Motions along the negative arc $-\infty < \varphi < 0$. This kind of motion is *unstable* (P^+) and *unstable*(P^-), since the moving point tends to the rest point as $t \to \infty$.
All the rest of the trajectories remain the same as in (9), since along them

$$\Phi(\varphi, \theta) \neq 0 . \tag{30}$$

These trajectories are everywhere dense on the torus and, therefore, *stable* (P) *in both directions.*

However the motions along the trajectories are *non-uniform*, the velocity being

$$\Phi(\varphi, \theta)\sqrt{1 + \alpha^2} , \tag{31}$$

and hence the motion is retarded on passing near (0,0).

8.4 *It is not always possible to orient a field of elements in the plane.*

Definitions: Consider a system of type

$$\frac{dx}{dt} = f_i(x_1, x_2, \ldots , x_n) , \tag{32}$$

$i = 1, 2, \ldots, n$, and systems in the so-called *symmetric form*:

$$\frac{dx_1}{X_1} = \frac{dx_2}{X_2} = \ldots = \frac{dx_n}{X_n}, \tag{33}$$

$$X_i = X_i(x_1, x_2, \ldots, x_n), \; i = 1, 2, \ldots, n .$$

System (32) assigns to each point $p = p(x_1, x_2, \ldots, x_n)$ a *vector* $(f_1, f_2, \ldots, f_n)$ whereas system (33) assigns to each point *a linear element* (line position): $dx_1 : dx_2 : \ldots : dx_n = X_1 : X_2 : \ldots : X_n$. The linear element is associated with *two* vectors, the vector $(X_1, X_2, \ldots, X_n)$ and the vector $(-X_1, -X_2, \ldots, -X_n)$.

Example (**55**, p. 36–37):

Consider *the field of linear elements* in the plane defined by equation

$$\frac{dy}{dx} = \cot\frac{\varphi}{2}, \tag{34}$$

where φ is the polar angle.

Consider equation

$$\frac{dx}{dy} = \tan\frac{\varphi}{2}, \tag{34$'$}$$

in the neighborhood of a point near which the absolute value of the right-hand member (of (34)) is *not* bonded.

It is obvious that the field of linear elements is defined and is continuous everywhere except at the point (0,0).

Introducing polar coordinates, we obtain

$$\cos\frac{3\varphi}{2}dr = r\sin\frac{3\varphi}{2}d\varphi . \tag{35}$$

Solving this equation, we obtain *three* integral half-lines

$$\varphi = \frac{\pi}{3}, \; \varphi = \pi, \; \varphi = \frac{5\pi}{3}, \; r > 0, \tag{36}$$

and *three* families of the similar curves (Figure 8.2)

$$r = \frac{a}{(\cos \frac{3\varphi}{2})^{\frac{2}{3}}} \tag{37}$$

with the parameter a and

(i) $-\frac{\pi}{3} < \varphi < \frac{\pi}{3}$,

(ii) $\frac{\pi}{3} < \varphi < \pi$,

(iii) $\pi < \varphi < \frac{5\pi}{3}$.

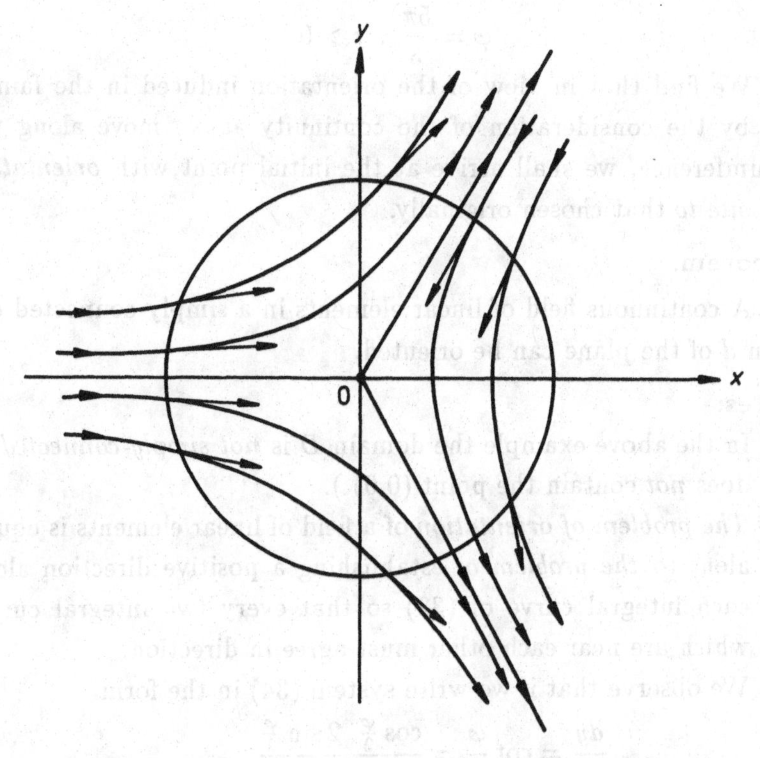

Fig. 8.2

The above field of elements *cannot be oriented.* In fact, choose the direction away from the origin on the half-line

$$\varphi = \frac{\pi}{3}\ , \ r > 0 \tag{38}$$

as positive.

Take a point p on this half-line and draw a circle through p with the center at (0,0). As we move along the circumference, say in the counterclockwise direction, considerations of continuity will asign as positive the direction toward the origin on the half-line

$$\varphi = \pi\ , \ r > 0 \tag{39}$$

and the direction away from the origin on the half-line

$$\varphi = \frac{5\pi}{3}\ , \ r > 0\ . \tag{40}$$

We find that in view of the orientation induced in the family (iii) by the consideration of the continuity as we move along the circumference, we shall arrive at the initial point with *orientation opposite to* that chosen originally.

Theorem.

A continuous field of linear elements in a simply-connected domain d of the plane can be oriented.

Notes:

(i) In the above example the domain D is *not simply-connected* (it does *not* contain the point (0,0)).

(ii) *The problem of orientation* of a field of linear elements is equivalent to *the problem of* establishing a positive direction along each integral curve of (33) so that every two integral curves which are near each other must agree in direction.

(iii) We observe that if we write system (34) in the form

$$\frac{dy}{dx} = \cot\frac{\varphi}{2} = \frac{\cos\frac{\varphi}{2}\ 2\sin\frac{\varphi}{2}}{\sin\frac{\varphi}{2}\ 2\sin\frac{\varphi}{2}}$$

$$= \frac{\sin\varphi}{2\sin^2\frac{\varphi}{2}} = \frac{\sin\varphi}{1-\cos\varphi} = \frac{r\sin\varphi}{r - r\cos\varphi}$$

or

$$\frac{dy}{dx} = \frac{y}{\sqrt{x^2 + y^2} - x} \tag{34)'$$

and then replace it by the system

$$\left.\begin{aligned} \frac{dx}{dt} &= \sqrt{x^2 + y^2} - x \\ \frac{dy}{dt} &= y \end{aligned}\right\}, \tag{34)''$$

then we introduce new singular points (points of equilibrium) filling the positive half of the axis.

(iv) The *selected positive direction* for the linear element through a point q: $q = q(x, y)$ on the circle may be indicated by a tangent unit vector $v = v(q)$. The angle

$$(2n\pi \ , \text{ or } \ 2n\pi + \pi) \tag{41}$$

through which v rotates as q spans the circle once in the positive sense, is a property of our field of directions.

The rotation angle for the circle in our example above is $-\pi$.

In this discussion the circle would be replaced by any other simply-connected curve containing the origin.

9. INTEGRAL EQUATIONS

9.1 *The integral equation:* $\varphi(x) = \lambda \int_0^\infty \sin(x\xi)\varphi(\xi)d\xi$ *possesses infinitely many linearly independent solutions for* $\lambda = \pm\sqrt{\frac{2}{\pi}}$.

Consider integral equation (**59**, p. 38):

$$\varphi(x) = \lambda \int_0^\infty \sin(x\xi)\varphi(\xi)d\xi \tag{1}$$

This equation possesses infinitely many linearly independent solutions for

$$\lambda = \pm\sqrt{\frac{2}{\pi}} \tag{2}$$

since for these values of λ the functions

$$\varphi = \varphi(x) = \sqrt{\frac{2}{\pi}} e^{-ax} \pm \frac{x}{a^2 + x^2} \tag{3}$$

satisfy equation (1) for arbitrary $a > 0$.

9.2 *A circle:* $|\lambda| < R$, *in which the eigenvalues* λ *of the integral equation:* $y(P) = \lambda \int K(P,Q)y(Q)dQ + f(P)$ *cannot have a limit.*

Preliminaries (**59**, p. 1–38):

(i) *Integral equations* are equations that contain the unknown function under the integral sign.

(ii) *Linear integral equations* are only equations of the form

$$a(x)\varphi(x) + f(x) = \int_a^b K(x,\xi)\varphi(\xi)d\xi \ , \tag{4}$$

where $x, \xi \in (a,b)$.

(iii) *Fredholm integral equations* (or *integral equations of the second kind*) are linear integral equations of the form

$$\varphi(x) = \int_a^b K(x,\xi)\varphi(\xi)d\xi + f(x) . \tag{5}$$

Note: Equation (5) is *homogeneous* if $f(x) = 0$.

(iv) *Integral equations of the first kind* are linear integral equations of the form

$$f(x) = \int_a^b K(x,\xi)\varphi(\xi)d\xi . \tag{6}$$

Note: The function $K = K(x,\xi)$ is the *kernel* of the integral equation and φ the unknown function. A kernel is *degenerate* if

$$K(P,Q) = \sum_{i=1}^{m} a_i(P)b_i(Q) , \tag{7}$$

where $a_i = a_i(P)$, $b_i = b_i(Q)$ are uniformly continuous in the bounded domain G.

Besides assume that $y(P)$ and $f(P)$ are uniformly continuous in G. Finally $a_i(P)$ and $b_i(Q)$ form linearly independent sets.

(v) Consider the integral equation

$$y(P) = \lambda \int K(P,Q)y(Q)dQ + f(P) , \tag{8}$$

where

$$\begin{aligned} K(P,Q) &= \sum_{i=1}^{m} a_i(P)b_i(Q) + K_1(P,Q) \\ &= A(P,Q) + K_1(P,Q) \end{aligned}$$

with $a_i(P)$, $b_i(Q)$, $K_1(P,Q)$, and $f(P)$ uniformly continuous in a bounded domain G.

Example. (**59**, p.27): To find *a circle:* $|\lambda| < R$ *in which the eigenvalues* λ *of the equation (8) cannot have a limit.* We work as follows:

In fact, this circle is given by the relation

$$|\lambda| < \frac{1}{M_1 D}\ , \tag{9}$$

where $|K_1(P,Q)| < M_1$, and $D:$ = Volume of region G.

Indeed, the solution of equation (8) is of the form

$$y(P) = \lambda \int \Gamma(P,Q,\lambda) f(Q) dQ + f(P)\ , \tag{10}$$

with

$$\Gamma(P,Q,\lambda) = \sum_{k=1}^{\infty} \lambda^{k-1} K^{(k)}(p,Q)\ ,$$

the resolvent of equation (8).

Replacing K by $A + K_1$ equation (8) takes the form

$$(E - \lambda A - \lambda K_1) y = f\ , \tag{11}$$

where E is the operator that transforms every function $y(P)$ into itself (i.e.: $Ey = y$ for every function $y(P)$), and A, K_1 operators with kernels $A(P,Q)$, $K_1(P,Q)$, respectively. Then

$$(E - \lambda K_1) y = \lambda A y + f\ (= \eta)\ . \tag{11$'$}$$

Since (9) holds, it follows from (10) that

$$y = \eta + \lambda \Gamma \eta \quad = (E + \lambda\Gamma)\eta\ , \tag{12}$$

where Γ is an operator corresponding to the resolvent $\Gamma(P,Q,\lambda)$ of the kernel $K_1(P,Q)$.

If the expression (12) for $y(P)$ is inserted in $(11)'$ we obtain

$$\eta = \lambda A(E + \lambda\Gamma)\eta + f$$

or

$$[E - \lambda A(E + \lambda\Gamma)]\eta = f \ . \tag{13}$$

Set

$$A \circ \Gamma(P, Q) = \int A(P, S)\Gamma(S, Q)dS \ . \tag{14}$$

We call the kernel

$$B(P, Q) = A \circ \Gamma(P, Q)$$

the symbolic product or *convolution of the kernels* $A(P, Q)$ and Γ (P, Q).

We note that the kernel $A(P, Q) + A \circ \lambda\Gamma(P, Q)$ of (13) is *degenerate* (see; relation (7)).

Note: The operators $A\Gamma$ is defined by the kernel $A \circ \Gamma$ if A and Γ are operators of the form

$$\psi(P) = \int K(P, Q)y(Q)dQ \tag{15}$$

(i.e.: symbolically $\psi = Ky$) with the kernels $A(P, Q)$ and $\Gamma(P, Q)$.

Suppose (7) and equation

$$\varphi(P) = \int_G K(P, Q)\varphi(Q)dQ + f(P) \ , \tag{16}$$

where $P, Q \in G$. Then

$$\varphi(P) = \sum_{i=1}^{m} a_i(P)c_i + f(P) \ , \tag{17}$$

where

$$c_i = \int_G b_i(Q)\varphi(Q)dQ \ . \tag{18}$$

Therefore

$$c_i = \int b_i(Q) \left[\sum_{j=1}^{m} a_j(Q)c_j + f(Q) \right] dQ$$

or

$$c_i = \sum_{j=1}^{m} K_{ij} c_j + f_i \ , \tag{19}$$

$i = 1, 2, \ldots, m$, where

$$K_{ij} = \int_G b_i(Q) a_j(Q) dQ \ ,$$

and

$$f_i = \int_G b_i(Q) f(Q) dQ \ .$$

The coefficient determinant $D(\lambda)$ of (19) is

$$D(\lambda) = \begin{vmatrix} 1 - K_{11} & -K_{12} & \ldots & -K_{1m} \\ -K_{21} & 1 - K_{22} & \ldots & -K_{2m} \\ \ldots & \ldots & \ldots & \ldots \\ -K_{m1} & -K_{m2} & \ldots & 1 - K_{mm} \end{vmatrix} .$$

Since the function $\Gamma(P, Q, \lambda)$ is a regular function of λ *in the circle*: $|\lambda| < R$, where

$$R = \frac{1}{M_1 D} \ , \tag{20}$$

the corresponding determinant for *the degenerate equation* (13) is also a regular function of λ in this circle. Since this determinant is 1 for $\lambda = 0$, it cannot be identically zero. Consequently its roots cannot have a limit point in this circle.

9.3 *A class of the integral equations with no eigenvalues.*

Definition (**55**, p. 40–42):

Integral equations of the form (8) that satisfy the following conditions are called *Volterra integral equations*

(i) Each coordinate of the points P, Q takes on every value between 0 and a constant $d > 0$.

(ii) $K(P,Q) = 0$ if at least one coordinate of the point Q is greater than the corresponding coordinate of the point P.

The 1-dimensional Volterra integral equation is one of the form

$$y(x) = \lambda \int_0^x K(x,\xi)y(\xi)d\xi + f(x) . \tag{21}$$

Example (**59**, p. 40–42).

Consider *the class of 1-dimensional Volterra integral equations* (21). Set

$$\overline{K}(x,\xi) = \begin{cases} K(x,\xi)|x-\xi|^{\epsilon} & , \quad 0 \le \xi \le x \\ 0 & , \quad \xi \ge x , \end{cases}$$

where $0 < \epsilon < 1$, which is uniformly continuous in the square: $0 \le x \le d$, $0 \le \xi \le d$.

Claim that the corresponding homogeneous equation

$$y(x) = \lambda \int_0^x K(x,\xi)y(\xi)d\xi \tag{22}$$

has *only trivial solutions in the class of the continuous functions of* x *for* $0 \le x \le d$ *and every* λ.

In fact, denote

Z: = the largest value of $|y(x)|$ for $0 \le x \le d$,

M: = the largest value of $|K(x,\xi)|$ for $0 \le x \le d$, $0 \le \xi \le x$.

Then we obtain from (22) that

$$|y(x)| \le \frac{|\lambda|^k}{k!} M^k d^k Z , \tag{23}$$

$k = 1, 2, \ldots$

But the last expression tends to 0 as $k \to \infty$. Consequently

$$y(x) \equiv 0 \tag{24}$$

on the interval, completing the proof of above claim.

Seek now a solution of (21) in the form of a power series

$$y(x) = y_0(x) + \lambda y_1(x) + \lambda^2 y_2(x) + \dots \quad (25)$$

Then we must have

$$y_0(x) = f(x) , \quad (26)$$

$$y_{k+1}(x) = \int_0^x K(x,\xi) y_k(\xi) d\xi , \quad (27)$$

$k = 0, 1, 2, \dots$

If $N{:} =$ the largest value of $|f(x)|$ in $(0, d)$, we get

$$|y_k(x)| \le \frac{M^k}{k!} d^k N , \quad (28)$$

$k = 0, 1, 2, \dots$

From this relation it is clear that *the series* (25) *is uniformly convergent in* λ *and* x for λ in an arbitrary large circle and $0 \le x \le d$.

Claim that *the Volterra equation has no eigenvalues.* In fact, consider the following system of linear algebraic equations in $0 \le x \le d$:

$$\begin{aligned}
f_1 &= y_1 - \lambda K_{11} y_1 \Delta\xi , \\
f_2 &= -\lambda K_{21} y_1 \Delta\xi + y_2 - \lambda K_{22} y_2 \Delta\xi , \\
f_3 &= -\lambda K_{31} y_1 \Delta\xi - \lambda K_{32} y_2 \Delta\xi + y_3 - \lambda K_{33} y_3 \Delta\xi , \\
&\dots\dots\dots\dots\dots\dots\dots\dots\dots\dots .
\end{aligned}$$

The above system can be successively solved for arbitrary but fixed λ, provided $|\Delta\xi|$ is sufficiently small, which is what we wish to establish.

Indeeed, one easily shows that for $\Delta x \to 0$ the solution of this system really approaches the solution of Volterra equation (21).

The determinant of the said system is

$$D = (1 - \lambda K_{11}\Delta\xi)(1 - \lambda K_{22}\Delta\xi)\ldots(1 - \lambda K_{nn}\Delta\xi) ,$$

where

$$\Delta x = \Delta\xi = \frac{d}{n} . \tag{29}$$

Therefore

$$d \geq (1 - |\lambda| M \Delta\xi)^{\frac{d}{\Delta\xi}} . \tag{30}$$

The right hand side of (30) is $\neq 0$ for $\Delta\xi$ sufficiently small, and increases for decreasimg $\Delta\xi$. As $\Delta\xi \to 0$ the right hand side of (30) tends toward

$$e^{-|\lambda| d M} .$$

The algebraic reason for the fact that the Volterra equation (21) has *no* eigenvalues is that

$$D \neq 0$$

and

$$D \not\to 0 \quad \text{as} \quad \Delta\xi \to 0 .$$

10. OTHER TOPICS

10.1 *A real nonsingular matrix with no real logarithm.*

Consider a real nonsingular matrix

$$C = -1 \tag{1}$$

with one row and column (**12**, p. 81).

Definition: A matrix C *has a real logarithm* means that there exists a real B such that

$$e^B = C\ . \tag{2}$$

Taking C according to (1) there need exist *no* real B such that (2) holds.

Note: However, for any real nonsingular matrix C implies that C^2 has a real logarithm.

10.2 *A differential equation with no single-valued solutions.*

Preliminaries (**12**, p. 108). Suppose that $A = A(z)$ is an $n \times n$ complex matrix, which is analytic and single-valued in the domain $D : 0 < |z - z_0| < a$, where a is some positive constant. Domain D is *not* simply connected, and thus the solutions of

$$\frac{dw}{dz} = A(z)w \tag{3}$$

need *not* be single valued; z is complex.

Example: Consider equation

$$\frac{dw}{dz} = \frac{1}{2z}w\ , \tag{3$'$}$$

where w is one-dimensional.

Then

$$\frac{d}{dz}\left(wz^{-\frac{1}{2}}\right) = 0 \qquad (4)$$

or

$$w = cz^{\frac{1}{2}}, \qquad (4)'$$

where C is a contant.

Therefore the solution, except for the case $c = 0$, is *not* single-valued in $0 < |z| < a$.

Note: The term *oequatio differentialis* (or *differential equation*) was first used by Leibniz (in 1676) to denote a relationship between the differentials dx and dy of two variables x and y (**30**, p. 3).

10.3 *A non-diferentiable subharmonic function*

Consider the function (**42**, p. 376)

$$u = u(x_1, x_2, x_3) = |x_1| \qquad (5)$$

in the parallelepiped

$$Q = \{(x_1, x_2, x_3) : -1 \le x_1 \le 1\,, -1 \le x_2 \le 1\,, -1 \le x_3 \le 1\}\,.$$

This function (5) is subharmonic in Q but has *no* derivative u_{x_1} at the points of the plane $x_1 = 0$.

Theorem: For the function $u = u(X)$ of the class $\mathbf{C}^2$ (in a domain D) to be *subharmonic* (in D) it is necessary and sufficient that at every point of D the Laplacian satisfies: $\Delta u \le 0$.

10.4 *An Initial Value Problem with continuity of first derivatives of initial data not yielding continuity of the first derivatives of the solution.*

Consider the initial value problem for the wave equation in three dimensions (**14**, p. 674):

$$u_{xx} + u_{yy} + u_{zz} - u_{tt} = 0 \tag{6}$$

with initial values

$$\left.\begin{aligned} u(0,x,y,z) &= 0 \\ u_t(0,x,y,z) &= f(r) \end{aligned}\right\} , \tag{7}$$

where

$$f(r) = \begin{cases} (1-r^2)^{\frac{3}{2}} & , \quad r^2 \leq 1 \\ 0 & , \quad r^2 \geq 1 \end{cases}$$

and

$$r^2 = x^2 + y^2 + z^2 .$$

These initial values imply *continuity of the initial derivatives up to the second order*

The explicit solution of the Problem (6)–(7) is given along the t-axis by

$$u(t,0,0,0) = \begin{cases} t(1-t^2)^{\frac{3}{2}} & , \quad t \leq 1 \\ 0 & , \quad t > 1 . \end{cases} \tag{8}$$

We have the relations:

$$u_t = 0 \ , \ u_{tt} = 0 \tag{9}$$

for $t > 1$ and

$$\left.\begin{aligned} u_t &= (1-t^2)^{\frac{3}{2}} - 3t^2(1-t^2)^{\frac{1}{2}} \\ u_{tt} &= -3(1-t^2)^{\frac{1}{2}} - 6t(1-t^2)^{1/2} + 3t^2(1-t^2)^{-1/2} \end{aligned}\right\} \tag{10}$$

for $t < 1$.

It is clear that u_t is continuous (for $t = 1$), but u_{tt} is not. If we set

$$w = u_t + u_x \ , \ v^1 = -u_y \ , \ v^2 = -u_z \ , \tag{11}$$

then w, v^1, v^2 satisfy the relations

$$\left.\begin{aligned} w_t - w_x + v^1_y + v^2_z &= 0 \\ v^1_t + v^1_x + w_y &= 0 \\ v^2_t + v^2_x + w_z &= 0 \end{aligned}\right\} \tag{12}$$

with initial conditions

$$w(0,x,y) = \begin{cases} (1-r^2)^{\frac{3}{2}} & , \quad r^2 \le 1 \\ 0 & , \quad r^2 > 1 \end{cases} \tag{13}$$

and

$$v^1(0,x,y) = 0 \ , \ v^2(0,x,y) = 0 \ . \tag{14}$$

Here w is continuous (for $t = 1$), but w_t is not.

Therefore continuity of first derivatives of the initial data does *not* guarantee continuity of first derivatives of the solution.

10.5 *A Cauchy sequence in the normed space of the continuous complex-valued functions f under norm $\|f\| = \sqrt{\int_0^1 |f(x)|^2 dx}$ failing to converge in this space.*

Consider the normed space $H(0,1)$ of continuous complex-valued functions f under norm

$$\|f\| = \sqrt{\int_0^1 |f(x)|^2 dx} \ . \tag{15}$$

Take (**17**, p. 74–75):

$$f_n = f_n(x) = \begin{cases} 0 & , \quad 0 \le x \le \frac{1}{2} - \frac{1}{10^n} \\ \frac{1}{2} + \frac{1}{2}10^n(x - \frac{1}{2}) & , \quad \frac{1}{2} - \frac{1}{10^n} \le x \le \frac{1}{2} + \frac{1}{10^n} \\ 1 & , \quad \frac{1}{2} + \frac{1}{10^n} \le x \le 1 \ , \end{cases} \tag{16}$$

$n = 1, 2, \ldots$

Each of these functions f_n clearly belongs to $H(0,1)$, and the sequence $\{f_n\}$ is obviously seen to be *convergent in norm.*

However, there does *not exist* a function $f \in H(0,1)$ such that

$$\lim_{n\to\infty} \|f - f_n\| = 0 \; . \tag{17}$$

In fact, assume there were such an f. Then from

$$\lim_{n\to\infty} \int_0^1 |f - f_n|^2 dx = 0 \; , \tag{18}$$

there would follow

$$\lim_{n\to\infty} \int_0^c |f - f_n|^2 dx = 0 \; , \tag{19}$$

where c is any positive constsnt less then $\frac{1}{2}$.

Now form (16) we see that except perhaps for a finite number of values of n,

$$f_n \equiv 0 \tag{20}$$

for all $x \in [0, c]$, so that

$$\int_0^c |f|^2 dx = 0 \tag{21}$$

must hold.

Since $f \in H(0,1)$ then from formula (21) we get

$$f \equiv 0 \tag{22}$$

for all $x \in [0, c]$, and hence $f \equiv 0$ for all $x \in [0, \frac{1}{2})$.

Similarly we conclude that $f \equiv 1$ for all $x \in (\frac{1}{2}, 1]$. Therefore f *cannot be continuous at* $x = \frac{1}{2}$, which is a contradiction.

10.6 *A functional $J[f(x)]$ attaining a weak minimum on the curve $f_0 = f_0(x)$ and not a strong minimum.*

Consider *the functional* (**39**, p. 51):

$$J[f(x)] = \int_0^\pi f^2(x)\left[1 - (f'(x))^2\right] dx \tag{23}$$

in the space of functions $f \in \mathbf{C}[0, \pi]$ that satisfy

$$f(0) = f(\pi) = 0 \ . \tag{24}$$

Definition: A functional $J[f(x)]$ attains *a weak* (or *strong*) *minimum on a curve* $f_0 = f_0(x)$ if for all admissible curves $f = f(x)$, located in an ϵ-neighborhood of the first (or zero-th) order of the curve $f_0 = f_0(x)$, we have

$$J[f(x)] \geq J[f_0(x)] \ (\text{or } J[f(x)] \leq J[f_0(x)]) \ , \tag{25}$$

where *ϵ-neighborhood of the n-th order of the curve* $f_0 = f_0(x)$, $0 \leq x \leq \pi$, means the collection of the curves $y = f(x)$ whose

$$\rho_n = \rho_n[f(x), f_0(x)] < \epsilon \ , \tag{26}$$

$n = 0, 1, 2, \ldots$, such that

$$\rho_n = \max_{0 \leq k \leq n} \ \max_{0 \leq x \leq \pi} |f^{(k)}(x) - f_0^{(k)}(x)|$$

is *the n-th order distance between the curves* $f = f(x)$ and $f_0 = f_0(x)$.

Claim that functional (23) attains a weak minimum on the curve: $y = 0$ and not a strong minimum.

In fact, for $y = 0$ we have $J = 0$, whereas for the curves, located in a first-order ϵ-neighborhood of the interval, where ϵ is any positive number (< 1), we have

$$|f'(x)| < 1 \ , \tag{27}$$

so that the integrand is positive for $f(x) \neq 0$ and, hence, the functional vanishes only when $y = 0$.

Therefore a weak minimum is attained on the function $y = 0$.

To show that a strong minimum is not attained it suffices to take

$$f(x) = \frac{1}{\sqrt{n}} \sin(nx) \; . \tag{28}$$

Then

$$\begin{aligned} J[f(x)] &= \frac{1}{n} \int_0^{\pi} \sin^2(nx)(1 - n\cos^2(nx))dx \\ &= \frac{1}{n} \int_0^{\pi} \sin^2(nx)dx - \frac{1}{4} \int_0^{\pi} \sin^2(2nx)dx \\ &= \frac{\pi}{2n} - \frac{\pi}{8} \end{aligned}$$

and for n sufficiently large for our curves

$$J < 0 \; . \tag{29}$$

On the other hand, for n sufficiently large, all the curves lie within an arbitrarily small neighborhood of zero-th order of the curve $y = 0$.

Therefore a strong minimum is *not* attained for $y = 0$.

11. OPEN PROBLEMS

11.1 Improperly posed problems of great potential importance in numerical analysis so far have not been reached by the main stream of active research.

How to improve and extend the computed solution is a question of high demand and requires more attention.

11.2 The existence problem for a regular transonic flow around given general profiles with given velocity at ∞ is difficult to be solved completely.

11.3 The difficulty of the correct statement of the problems for equations of mixed type in higher dimensions still remains.

11.4 It would be interesting to clear up the questions whether there is an extremal principle for the boundary value problems of mixed type.

11.5 To consider regions of mixed type (elliptic-parabolic-hyperbolic) multi-connected with arbitrary parabolic curves is difficult to handle.

11.6 The study of higher order equations and systems of mixed type equations is a question of great interest despite the little mathematical progress.

12. REFERENCES

1. **Atkinson, K.**: "Elementary Numerical Analysis", John Wiley. & Sons, 1985, 288–292.
2. **Bakusinskii, A. B.**: "The Solution by Difference Methods of an Ill-posed Cauchy Problem for a Second Order Abstract Differential Equation" (Russian), Differ. Uravn., 8, 1972, 881–890.
3. **Bakusinskii, A. B.**:"Difference Methods of Solving Ill-posed Cauchy Problems for Evolution Equations in a Complex Banach Space" (Russian), Differ. Uravn., 8, 1972, 1661–1668.
4. **Barros-Neto, J.**: "Remarks on Non-existence of Solutions for an Abstract Differential Equation", An. Acad. Sci. Brasil Ci, 38, 1966, 1–4.
5. **Birkhoff, G.**, and **Rota, Gian-Carlo**: "Ordinary Differential Equations", 3rd Ed., John Wiley & Sons, Inc., 1978, p. 8, 20–23, 97, 100, 106–107, 114, 117, 121–122, 142, 152, 157.
6. **Bitsadze, A. V.**: "Equations of the Mixed Type" (translated in English by P. Zador), The MacMillan Company, New York, 1964, 38–40.
7. **Bitsadze, A. V.**, and **Kalinichenko, D. F.**: "A Collection of Problems on the Equations of Mathematical Physics", Transl. in English by I. M. Volosov, and I. G. Volosova, MIR Publishers, Moscow, 1980, p. 46, 63–82, 90–92, 109, 202–203, 224–231, 243–258.
8. **Bloom, F.**: "Ill-posed Problems for Integro-differential Equations in mechanics and Electromagnetic Theory", SIAM Studies in Applied Math., Philad., 1981, p. 21–28, 88–89, 213.
9. **Broman, A.**: "On Pathological Properties of Partial Differential Equations", Colloquium, Univ. of Victoria, 1971, 25–33.
10. **Carraso, A.**, and **Stone, A. P.** (Editors): "Improperly

Posed Boundary Value Problems", Pitman "Research Notes in Mathematics", London, 1975, p. 1–15.

11. **Chernoff, P. R.**: "Two Counter-examples in Semi-group Theory in Hilbert Space", Proc. Amer. Math. Soc., 56, 1976, 253–255.
12. **Coddington, E. A.**, and **Levinson, N.**: "Theory of Ordinary Differential Equations", McGraw-Hill Book Co., 1955, p. 3, 6, 14, 19, 41, 45, 53–54, 65, 69–70, 81, 108, 371–378, 381.
13. **Cohen, P.**: "The Non-uniqueness of the Cauchy Problem", Techn. Report No. 93, Applied Math. & Stat. Laboratory, Stanford University, 1960, p. 383.
14. **Courant, R.**, and **Hilbert, D.**: "Methods of Mathematical Physics", II, Interscience Publ., 1962, p. 14, 148–151 231, 240–244, 280–282, 303–305, 369, 486, 674.
15. **Derrick, W. R.**, and **Grossman, S. I.**: "Elementary Differential Equations with Applications", 2nd Ed., Addison-Wesley Publ. Co., 1981, p. 11.
16. **Duchateau, P.**, and **Zachmann, D. W.**: "Partial Differential Equations", Schaum's Outline Series, McGraw-Hill Book Co., 1986, p. 13–14, 20, 71.
17. **Epstein, B.**: "Partial Differential Equations", Robert E. Krieger Publ. Co., New York, 1975, p. 1–4, 17–24, 42–54, 74–75, 138–139, 198, 253–255.
18. **Fattorini, H. O.**: "The Cauchy Problem", Addison-Wesley Publ. Co., London, Vol. 18, 1983, p. 6–11, 29–30, 39–49, 112–114, 341–354.
19. **Fefferman, CH. L.**: "The Uncertainty Principle", Bulletin of the Amer. Math. Soc., Vol. 9, No. 2, 1983, 138–139.
20. **Finizio, N.**, and **Ladas, G.**: "An Introduction to Differential Equations", Wadsworth Publ. Co., Belmont, 1982, p. 13.
21. **Foguel, S. R.**: "A Counter-example to a Problem of Sz.-Nagy", Proc. Amer. Math. Soc. 15, 1964, 788–790.

22. **Fraleigh, J. B.**: " Calculus with Analytic Geometry", Addison-Wesley, Publ. Co., 1st Ed., 1982, p. 724.
23. **Garabedian, P. R.**: "Partial Differential Equations", John Wiley & Sons, Inc., New York, 1964, p. 101–102, 108, 110, 120, 176, 236, 450–455, 456.
24. **Goldstein, J. A.**: "Some Counter-examples involving Self-adjoint Operators", Rocky Mountains J. Math., 2, 1972, 143–149.
25. **Hajek, O.**: Book Review of "Differential Systems involving impulses, by S. G. Pandit, and S. G. Deo", Bulletin of the Amer. Math. Soc., Vol. 12, No. 2, 1985, p. 273.
26. **Hajek, O.**: "Dynamical Systems in the Plane", Academic Press, London, 1968, p. 25–26, 31–32, 50, 56–59.
27. **Hale, J. K.**: "Ordinary Differential Equations", Wiley-Interscience, London, 1969, p. 46–49, 107–121.
28. **Hamel, G.**: "Eine Basis aller Zahlen und die unstetigen Lösungen der Funktionalgleichung $f(x+y) = f(x) + f(y)$", Math. Ann. 60, 1905, 459–462.
29. **Hartman, Ph.**: "Ordinary Differential Equations", John Wiley & Sons, Inc., New York, 1964, p. 8, 15-23, 40–41.
30. **Ince, E. L.**: "Ordinary Differential Equations", Dover Publ., Inc., New York, 1956, p. 3, 66–71.
31. **Ince, E. L.**, and **Sneddon, I. N.**: "The Solution of Ordinary Differential Equations", Longman Scientific & Technical, England, 13–18, 1987.
32. **Ivanov, V. K.**: "The Value of the Regularization parameter in Ill-posed Control Problems" (Russian), Differ. Uravn. 10, 1974, 2279–2285.
33. **Jakubov, S. JA.**: "Non-local Solvability of Boundary Value Problems for Quasi-linear Partial Differential Equations of Hyperbolic Type" (Russian), Dokl. Akad. Nauk SSSR, 176, 1967, 279–282.

34. **John. F.**: " A Note on Improper Problems for Partial Differential Equations", Comm. Pure Appl. Math., 8, 1955, 591–594.
35. **John, F.**: "Numerical Solution of Problems which are not Well-posed in the sense of Hadamard", Proc. Rome Symp. Prev. Int. Comp. Center, 1959, 103–116.
36. **Kaplan. W.**: "Elements of Differential Equations", Addison-Wesley Publ. Co., Inc., London, 1964, p. 210–213, 212–216,220–221, 235–237, 240–241, 246.
37. **Karasik, B. G.**: "Regularization of an Ill-posed Cauchy Problem for Operator-differential Equations of Arbitrary Order" (Russian), Izv. Akad. Nauk Azerbaidzan. SSR Ser. Fiz.-Tech. Mat. Nauk, 1976, 9–14.
38. **Knops, R. J.** (Editor): Symposium "On Non-well-posed Problems and Logarithmic Convexity", Lecture Notes in Mathematics, Vol. 316, Springer-Verlag, Berlin & N. York, 1973.
39. **Krasnov, M. I.**, **Makarenko, G.I.**, and **Kiselev, A. I.**: "Problems and Exercises in the Calculus of Variations", MIR Publishers, Moscow, 1975, p. 35, 51.
40. **Krein, S. G.**, and **Prozorovskaya, O. I.**: "Analytic Semi-groups and Incorrect Problems for Evolutionary Equations" (Russian), Dokl. Akad. Nauk SSSR 133, 1960, 277–280.
41. **Krein, S. G.**, and **Prozorovskaya, O. I.**: "Approximate Methods of Solving Ill-posed Problems" (Russian), Z. Vycisl. Mat. i Mat. Fiz 3, 1963, 120–130.
42. **Krzyzanski, M.**: "Partial Differential Equations of Second Order", PWN-Polish Scientific Publishers, Warsaw, 1971, p. 61–62, 125–126, 169–170, 176, 184, 215, 235–237, 266, 288–289, 370–376.
43. **Lavrentiev, M. M.**: "Some Improperly Posed problems of Mathematical Physics", Springer, New York, 1967.
44. **Lavrent'ev, M. M.**, **Romanov, V. G.**, and **Shishatskii, S. P.** (translated in English by J. R. Schulenberger): "Ill-posed

Problems of Mathematical Physics and Analysis", Translations of Mathematical Monographs, Vol. 64, Amer. Math. Soc., Providence, R.I., 1986, vi +290 p. (Book Review by R.J. Knops, Bulletin of the Amer. Math. Soc., Vol. 19, No. 1, 1988, 332–337).

45. **Levine, H. A.**: "Instability and Nonexistence of Global Solutions to Nonlinear Wave Equations of the form $Pu_{tt} = -Au + F(u)$", Trans. Amer. Math. Soc. 192, 1974, 1–21.
46. **Levine, H. A.**, and **Payne, L. E.**: "On the Nonexistence of Global Solutions to some Abstract Cauchy Problems of Standard and Nonstandard types", Rend. Mat. (6), 8, 1976, 413–428.
47. **Lewy, H.**: "An Example of a Smooth Linear Partial Differential Equation without Solution", Ann. of Math., (2)66, 1957, 155–158.
48. **Lions, J. L.**: "Sur la Stabilization de certains Problèmes mal posés" (English Summary), Rend. Sem. Mat. Fis. Milano, 36, 1966, 80–87.
49. **Ljubic, Ju. I.**: "Investigating the deficiency of the Abstract Cauchy Problem" (Russian), Dokl. Akad. Nauk SSSR, 166, 1966, 783–786.
50. **Magiros, D. G.**: "Stability Concepts of Dynamical Systems", Information and Control, Vol. 9, No. 5, 1966, 546–547.
51. **Masuda, K.**: "Anti-locality of the one-half power of Elliptic Differential Operators", Publ. Res. Inst. Math. Sci. 8, 1972, 207–210.
52. **Mlak, W.**: "Limitations and Dependence on Parameter of Solutions of Non-stationary Differential Operator Equations", Ann. Pol. Math. 6, 1959, 305–322.
53. **Morozov, V. A.**: "Linear and Nonlinear Ill-posed Problems" (Russian), Mathematical Analysis, Vol. 11, 129–278, 180, Akad.

Nauk SSSR Vsesojuz. Inst. Naucn. i Techn. Informacii, Moscow, 1973.

54. **Murata, M.**: "Anti-locality of certain Functions of the Laplace Operator", J. Math. Soc. Japan, 25, 1973, 556–564.
55. **Nemytskii, V. V.**, and **Stepanov, V. V.**: "Qualitative Theory of Differential Equations", Princeton, New Jersey, Princeton Univ. Press, 1972, p. 21–31, 35–38, 308–320, 325, 329–330, 331–332, 340–341, 343–347, 374, 380–400, 406–408, 410–411, 468, 485–486, 512–519.
56. **Odhnoff, J.**: "Un Example de Non-unicité d' une Équation Différentielle Opérationelle", C. R. Acad. Sci. Paris, 258, 1964, 1689–1691.
57. **Packel, E. W.**: "A Semigroup Analogue of Foguel's Counter-example", Proc. Amer. Math. Soc. 21, 1969, 240–244.
58. **Payne, L. E.**: "Improperly Posed Problems in Partial Differential Equations", CBMS Regional Conference Series, Vol. 22, Philadelphia, SIAM, 1975, 58–59.
59. **Petrovskii, I. G.**: "Lectures on the Theory of Integral Equations", Graylock Press, Rochester, N. Y., 1957, 1–42.
60. **Plis, A.**: "On Non-uniqueness in Cauchy problem for an Elliptic Differential Equation of the Second Order", Bull. Acad. Polon. Sc. 11, 1963, 95–100.
61. **Plis, A.**: "A Smooth Linear Elliptic Differential Equation without any Solution in a Sphere", Comm. Pure & Appl. Math., Vol. 14, No. 3, 1961, 599–617.
62. **Plis, A.**: "Non-uniqueness in Cauchy's Problem for Differential Equations of Elliptic Type", J. Math. & Mech., Vol. 9, 1960, 557–562.
63. **Polya, G.**, and **Szegö, G.**: "Problems and Theorems in Analysis", I, Springer-Verlag, 1972, p. 166 (Problem 315 §4 Harmonic Functions).
64. **Powers, D. L.**: "Elementary Differential Equations with Boun-

dary Value Problems", Prindle, Weber, & Schmidt, Boston, 1985, p. 49–50, 57–58.

65. **Pucci, C.**: "Sui Problemi di Cauchy Non bene posti", Rend. Accad. Naz. Lincei, 18, 1955, 473–477.
66. **Radmiz, A.**: "The Incomplete Cauchy problem in Banach Spaces", Ph. D. Dissertation, Univ. of Calif., L. A., 1970.
67. **Rassias, J. M.**: "Mixed Type Equations", BSB Teubner, Leipzig, 90, 1986.
68. **Rassias, J. M.**: "Mathematics-Space Technology", Athens, Greece, 1981.
69. **Rassias, J. M.**: "Lecture Notes on Mixed Type Partial Differential Equations", World Sci. Publ. Co. Pte Ltd, Singapore, 1990.
70. **Ross, SH. L.**: "Differential Equations", Blaisdell Publ. Co., London, 1965, 496–497.
71. **Sperb, R. P.**: "Maximum Principles and their Applications", Academic Press, Vol. 157, London, 1981, p. 8–18, 25–28, 94–96.
72. **Tikhonov, A. N.**, and **Arsenin, V. Y.**: "Solutions of Ill-posed Problems", V. H. Winston and Sons-Wiley, New York, 1977.
73. **Vemuri, V.**, and **Karplus, W. J.**: "Digital Computer Treatment of Partial Differential Equations", Prentice-Hall, Inc., Englewood Cliffs, 1981, p. 73.
74. **Vichnevetsky, R.**: "Computer Methods for Partial Differential Equations", Prentice-Hall, Inc., Engl. Cliffs, Vol. 1, 1981, p. 13.
75. **Wilf, H. S.**: "Mathematics for the Physical Sciences", Dover Publications, Inc., New York, 1962, p. 143-145, 150, 153.
76. **Wilson, H. K.**: "Ordinary Differential Equations", Addison-Wesley Publ. Co., London, 1971, p. 24–28, 255, 269–270, 295, 308–309, 315–317, 324–329, 367.
77. **Zaidman, S. D.**: "Un Teorema di Esistenza per un Problema

Non bene posto", Atti Accad. Naz. Lincei Rend. Cl. Sci. Fis. Mat. Natur. (8), 35, 1963, 17–22.

78. **Zill, D. G.**: "A First Course in Differential Equations with Applications", Prindle, Weber & Schmidt, 1y82, p. 8, 34, 36–37.

Note: Other References are mentioned through the Topics.

SUBJECT INDEX

Abstract Cauchy problem 4
Analyticity 36
Asymptotically attractive point 94
Asymptotically stable set 97

Banach space 36
Bendixson's nonexistence criterion 16

Cauchy conditions 19
Cauchy problem 17
Cauchy-Euler equation 135
Cauchy-Riemann equations 24
Cauchy-Schwarz inequality 42
Center 100, 102
Characteristics singularity 75
Characteristics 70
Closed path 15
Coefficient determinant 154
Compact space 120
Components 121
Connected space 120
Conservation law 76
Continuity 1
Continuum 60
Convergent sequence in norm 162
Convolution of kernels 153

De L' Hospital's Rule 4
Degenerate equation 154
Derivative of Lyapunov function 106
Differentiability 4
Differential system 14
Dirichlet problem 31, 80
Discontinuous function 3

Distance 96
Dynamical system 117, 138

ϵ-approximate solution 54
Eigenvalues 150, 152, 156
Elliptic system 80
ϵ-neighborhood of n-th order 163
Envelope 49, 134
Everywhere dense in 142
Exceptional points 132
Existence 10
Explosive solution 38
Exterior problem 72
Extrema 6

Field of linear elements 146
Fourier coefficients 23
Fredholm integral equations 151
Functional 163
Fundamental solutions 76

Global solution 11
Green's theorem 15, 40

Hadamard stability 129
Hadamard's Counter-example 32
Harmonic function 19
Heat equation 26
Heat diffusion equation 37
Homogeneous function 35

Implicit Function Theorem 136
Infinite continued fraction 124
Integral equations 150, 151

Invariant set 96

Jump relation 77

Kernel 151

Lagrange stability 118
Laplace equation 9, 19
Linear approximation 100
Linear elliptic equation 7
Linearity 1
Lipschitz condition 10, 14
Local solution 11
Lyapunov function 105, 106
Lyapunov stability 123

Matrix with real logarithm 158
Mean value equation 1
Metric space 117

Negative maximum 7
Neutrally stable point 94
Nonlinear Cauchy problem 20
Nonlinear elliptic equation 89
Nonlinear function 1
Nonlinear system 111
Nonperiodic solutions 140
Non-uniform motion 144
N-th order distance 163

ω-limit point 117
Open problems 165

p-discriminant locus 132, 134

Permissible solution 79
Perturbations 100, 112, 128
Perturbed linear 100
Phenomenon of conservation 77
Picard's theorem 10, 11
Poisson stability 124
Polar equations 112
Polygonal solutions 56
Positive definite function 106
Positive maximum 7
Potential 27
Proper node 109, 113

Regular at infinity 74
Regular family 138
Regular in exterior region 74
Resolvent of equation 152
Rotational symmetry 32

Schwarz reflection principle 19, 22, 24
Singular locus 132
Singularities 132
Spiral point 109, 113
Stability 94
Stability in the sense of Lyapunov 97
Stability in the sense of Poincaré 97
Stable point 94
Stable set 96
Strictly stable point 94
Strong maximum principle 6
Strong minimum on a curve 163
Subharmornic function 159
Successive approximations 54
Symbolic product 153

Systems in symmetric form 145

Temperature 35

Unbounded function 8
Uniqueness 46
Unstable point 94
Universe 35

Volterra integral equations 154, 155

Wave equation 20
Weak maximum principle 6
Weak minimum on a curve 163
Weak solution 71
Weierstrass formula 38
Well-posed problem 129

NAME INDEX

ARSENIN, V. Y. 172
ATKINSON, K. 166

BAKUSINSKII, A. B. 140
BARBASHIN, E. A. 140
BARROS-NETO, J. 166
BIRKHOFF, G. 166
BITSADZE, A. V. 166
BLOOM, F. 166
BROMAN, A. 166
BROWN, TH. 94, 95

CARASSO, A. 166
CHERNOFF, P. R. 167
CODDINGTON, E. A. 167
COHEN, P. 167
COURANT, R. 167

DEO, S. G. 168
DERRICK, W. R. 167
DUCHATEAU, P. 167

EPSTEIN, B. 167

FATTORINI, H. O. 167
FEFFERMAN, CH. L. 167
FINIZIO, N. 167
FOGUEL, S. R. 167
FRALEIGH, J. B. 168

GARABEDIAN, P. R. 168
GOLDSTEIN, J. A. 168
GROSSMAN, S. I. 167

HAJEK, O. 168
HALE, J. K. 168
HAMEL, G. 3, 168
HARTMAN, PH. 60, 168
HILBERT, D. 167

INCE, E. L. 168
IVANOV, V. K. 168

JAKUBOV, S. JA. 168
JOHN, F. 169

KALINICHENKO, D. F. 166
KAPLAN, W. 169
KARASIK, B. G. 169
KARPLUS, W. J. 172
KISELEV, A. I. 169
KNESER, H. 60
KNOPS, R. J. 169, 170
KRASNOV, M. I. 169
KREIN, S. G. 169
KRZYZAŃSKI, M. 169

LADAS, G. 167
LAVRENTIEV, M. M. 169
LEBESGUE, H. 27
LEIBNIZ, G. W. 159
LEVINE, H. A. 170
LEVINSON, N. 167
LEWY, H. 25, 170
LIONS, J. L. 170
LJUBIC, JU. I. 170

MAGIROS, D. G. 170

MAKARENKO, G. I. 169
MASSERA 105, 107
MASUDA, K. 170
MLAK, W. 170
MOROZOV, V. A. 170
MURATA, M. 171

NEMYTSKII, V. V. 171

ODHNOFF, J. 171

PACKEL, E. W. 171
PANDIT, S. G. 168
PAYNE, L. E. 170, 171
PERRON, O. 102, 103
PETROVSKII, I. G. 171
PLIS, A. 171
POLYA, G. 171
POWERS, D. L. 171
PROZOROVSKAYA, O. I. 169
PUCCI, C. 172

RADMIZ, A. 172
RASSIAS, J. M. 172
ROMANOV, V. G. 169
ROSS, SH. L. 172
ROTA, G. C. 166

SCHULENBERGER, J. G. 169
SHIFFMAN, M. 1
SHISHATSKII, S. P. 169
SNEDON, I. N. 168
SPERB, R. P. 172
STEPANOV, V. V. 171

STONE, A. P. 166
SZEGÖ, G. 171

TAYLOR, G. I. 131
TIKHONOV, A. N. 172

VEMURI, V. 172
VICHNEVETSKY, R. 172
VOLOSOV, I. M. 166
VOLOSOVA, I. G. 166

WEIERSTRASS, K. W. T. 33
WHITNEY, H. 138
WILF, H. S. 172
WILSON, H. K. 172

ZACHMANN, D. W. 167
ZADOR,P. 166
ZAIDMAN, S. D. 173
ZILL, D. G. 173